Innovation in TBM Traffic Tunnels

Tunnel Boring Machine (TBM) constructed tunnels are widespread, and can deliver significant environmental and cost benefits. However, as noted in the noteworthy examples of TBM traffic tunnels presented in this book, there are still important challenges associated with them, linked in particular to structural safety in the event of earthquakes, as well as cost and safety issues during operation.

To face these challenges, **Innovation in TBM Traffic Tunnels** presents three innovative concepts in the field of construction of TBM rail and road tunnels: the TISB concept that improves the structural safety of those built on soft soil in seismic areas, and the TMG and TMF concepts, for rail and road tunnels, respectively, that allow for significant reduction of their cost and the improvement of safety during operation. Examples of the application of these new concepts in the conceptual design of specific tunnel cases are presented and compared with solutions based on common approaches, demonstrating the additional benefits of these concepts. The book also draws attention to other innovations in TBM tunnelling that may improve the construction of tunnels in the future, especially when using the concepts mentioned above.

Innovation in TBM Traffic Tunnels is aimed at professionals involved in the planning, design, and construction of tunnels for transport infrastructure, including authorities, consultants and construction companies, worldwide.

Innovation in TBM Traffic Tunnels

Silvino Pompeu-Santos

CRC Press
Taylor & Francis Group
Boca Raton   London   New York   Leiden

CRC Press is an imprint of the
Taylor & Francis Group, an **informa** business

A BALKEMA BOOK

Designed cover image: Courtesy Dragages Hong Kong Limited, a member of Bouygues Construction

First published 2022
by CRC Press/Balkema
4 Park Square, Milton Park / Abingdon, Oxon, OX14 4RN
e-mail: enquiries@taylorandfrancis.com
www.routledge.com – www.taylorandfrancis.com

CRC Press/Balkema is an imprint of the Taylor & Francis Group, an informa business

© 2022 Silvino Pompeu-Santos

Library of Congress Cataloging-in-Publication Data
A catalog record has been requested for this book

ISBN: 978-0-367-70163-5 (hbk)
ISBN: 978-0-367-70164-2 (pbk)
ISBN: 978-1-003-14486-1 (ebk)

DOI: 10.1201/9781003144861

Typeset in Times New Roman
by MPS Limited, Chennai, India

Dedication

To my granddaughters, Sara and Clara,

and my grandson, David

Table of contents

About the Author

Born in 1950, Silvino Pompeu-Santos graduated in Civil Engineering from the University of Lisbon in 1974. In 1983 obtained his PhD degree and in 1995 became Research Director at LNEC (National Laboratory of Civil Engineering) in Lisbon. In 2013, he founded the consulting office SPS Consulting, of which he is the main partner and the Managing Director.

Professional activity developed in the area of structures, mainly concrete and masonry, and in the rehabilitation of heritage buildings. In recent years, expansion to other areas, namely tunnels (several patents at national and European level), and planning of transport infrastructures and spatial planning.

Author of hundreds of papers and reports, as well as dozens of lectures, and author/editor of several books on his fields of interest.

Chairman of the CIB (International Council for Construction) Commission Wall Structures between 1996 and 2014, he was awarded the CIB PC Commendation in 2006.

Vice President of IABSE (International Association for Bridge and Structural Engineering) between 1999 and 2007, in 2013 he was elected Honorary Member of IABSE.

Member of FIB (International Federation for Structural Concrete) and of FIB/TG1.4- Tunnels, of which he is the Convenor of WP1.4.2- Openings in Walls of Tunnels.

Ackowledgments

The author of this volume would like to express his sincere thanks to the following entities and personalities.

To Herrenknecht AG and its staff members Martin Foster, Fernando Molina and Konrad Fuchs for the exchanging of views on the construction of TBM traffic tunnels.

To Fogtec GmbH and its staff member Karl-Heinz Radl for the exchanging of views on the fire-fighting in traffic tunnels.

To Prof. Uwe Starossek from the Technical University of Hamburg-Harburg for the exchange of views and invitation to give the 2014 Annual Conference of the Structures Institute, entitled "Alternative TBM tunnel solution for the Fehmarnbelt Fixed Link".

To Prof. Bernd Hillemeier from the Technical University of Berlin for his friendship and exchange of views on the construction of tunnels.

To my old friend Carlos Garcia, Civil engineer, for the passionate conversations about the construction of tunnels around the world.

To my son, Rodolfo, orthopaedic surgeon and sports medicine specialist, for the advice on issues related to emergency procedures in tunnels in operation.

Special thanks to Prof. Philippe Van Bogaert from the University of Ghent, for his encouragement and collaboration in revising the text and authoring the Preface.

Last but not least, to Taylor and Francis and its staff members Tony Moore, Alistair Bright, Léon Bijnsdorp, Kaustav Ghosh and Lukas Goosen for their action in publishing the book.

Lisbon, September 2022
Silvino Pompeu-Santos

Foreword

The volume that you are about to read reflects the vast experience of the author in the field of making adequate choices for tunnel configurations. Throughout his professional career Silvino Pompeu-Santos has proved to be an excellent specialist in this activity, as well as a gifted presenter and debater.

In this book he takes the reader through the history of tunnelling right from the early days of Brunel and other experiments to the present state of the art. Concerning the latter, detailed descriptions are given of the characteristics and the background reasons for the construction of most of the world's famous TBM-bored tunnels. This includes both railway and road tunnels. Emphasis is on safety measures and equipment, both for the structure itself and for travellers.

Obviously, the core of the book are the author's own ideas on alternatives to classical tunnel arrangements, as he has developed during several years. These proposals, called TISB, TMG and TMF, are intended to improve specific characteristics of bored tunnels. The first aims at enhanced resistance to seismic effects or displacements due to soft soil sliding. The TMG and TMF systems provide answers to reducing cost and to allow the necessary technical systems, imperative to safe traveling.

The author does not limit the text to a plea in favour of his own ideas, but he extensively tests them on several practical cases, be it build or still in a preparatory phase. While doing this, he does not shy away from any aspect and carefully examines all elements of the design. In particular, and for each case, the alternatives conceived during the design are thoroughly analysed and the author's own design is weighed against all parameters that influence the project. Obviously, the author always decides that his proposal is the best, which of course we cannot deny him. Most importantly, however, he opens discussions and reflection, so that in a case that would concern him or her, the reader can take the reasoning to heart.

Reading chapters 9 and 10, the former about the immense Fehmarnsund-link and the latter on future projects as the Gibraltar Strait crossing, Maglev train connections and perhaps the Hyperloop system, may be the most attractive to read for enthusiasts of tunnel construction. For readers who are more focused on technical issues, these chapters give insight in the particular characteristics of the projects, their history and present state and how realistic they may have become. One can trust that some of them will be built one day and that society will benefit from new connections between countries and people. Because infrastructure serves to connect people and ideas, for trade and industry, but just as well for cultural and tourist exchange, in short, to make the world a place for every individual.

In this book the reader will not find design formulas, nor details of performed numerical simulations or geotechnical data and derivations. Those may be found in the already abundant specialized literature. Adversely the reader will find comments, discussion and considerations on making adequate choice, holistic views on structural, safety and environmental issues related to mechanized tunnelling. I think the latter may be of greater use than the former.

Albeit time was short, I have read this book with growing interest. Possibly the author was not so pleased with the many suggestions I made, although I think he has forgiven me. Those suggestions were merely intended as an appreciation of his work. This book may very well be of wider use than the author intended, for example as an introductory course in mechanized tunnelling for undergraduate students. It also may serve for a broader public to discover the exciting occupation of those involved in tunnelling. The examples being discussed clearly show that tunnel projects may very well fail because alternatives have not – or only shallowly – considered, mostly under time pressure.

As for each civil engineering project, discussions should be open and arguments thoroughly waged. Often and given time, those who helped deciding, regret some choices that were made. This proves once again that too little effort is given to the conceptual phase of structural design, whether it be tunnels or other structures. A book like Silvino Pompeu-Santos's "Innovation in TBM Traffic Tunnels" should be helpful in such stages of project development. In this sense it fills a gap in technical literature and deserves a prominent place as a handbook for conceptual design of tunnels.

Ghent, February 2021
Prof dr ir Philippe Van Bogaert
Civil Engineering Department
Ghent University
Belgium

Introduction

Tunnels, like bridges and viaducts, are basic infrastructure that have played a decisive role throughout Human History, facilitating trade and travel between places separated by mountains, rivers, bays, and other types of obstacles.

As transport infrastructure, tunnels have a relatively low environmental impact (less than viaducts and bridges, for example), contributing to a "greener" environment and a more sustainable future. In urban areas, when passing traffic to underground, reduce pollution at surface, thus contributing to the improvement of the quality of life of citizens.

Some attributes of tunnels, however, limit their wider application, such as the high costs of construction and the risks during operation. With regard to costs, while they are in fact generally higher than those of bridges, considering the lifetime costs, tunnel solutions can nevertheless be more cost-effective. As far as risks during the operation are concerned, measures are available today to reduce these risks to acceptable levels.

Despite the conservative nature of tunnel construction, as, indeed, the construction industry in general, it benefited from the ongoing technological revolution, adopting increasingly sophisticated equipment, processes and technologies, with improvements in both the project and execution of tunnels.

Thus, tunnel construction made a tremendous progress in the last decades, being today available a wide variety of reliable techniques, from improved traditional mining techniques and the cut-and-cover methods, to more sophisticated techniques, such as immersed tunnels and, in particular, TBM tunnels, that is, made with TBMs, in which practically all activities are incorporated, with great benefits in terms of construction agility and cost reduction.

However, when it comes to tunnel innovation, there are sometimes difficulties in its implementation, as there is not always a clear understanding between the stakeholders involved, namely in the triangle, owners, consultants and contractors.

In fact, unlike contractors, owners are often reluctant to innovation, because of fear of being the first to try something new. To some owners "tried and tested" means "safe and steady", innovation means "no idea if it will work or not".

Now, the truth is that innovation in tunneling is, in most cases, just refinements or new uses of well-known methods and technologies, which can get the job done faster, cheaper and safer. However, the right combination of existing ideas can lead to innovative and very cost-benefit solutions with benefits for all. Consultants have an important role to play here, convincing owners to believe in them and not be afraid to try them out.

TBM tunnels are today the most common and most promising technique for the construction of tunnels, with cost-effective solutions in almost all types of ground conditions. However, despite the great advances observed in recent times, TBM tunnelling still face significant challenges.

With regard TBM traffic tunnels, the challenges to be faced are, inter alia, improving the reliability of structural safety when built on soft ground in seismic areas, increasing safety during operation and further reducing construction costs.

DOI 10.1201/9781003144861-1

To face these challenges, three new concepts for TBM traffic tunnels (railway and road tunnels) have recently been developed by the author: the TISB ("Tunnel of Improved Seismic Behaviour") concept, the TMG ("Tunnel Multi Gallery") concept and the TMF concept ("Tunnel Multi Floor").

The TISB concept allows to improve the structural safety of TBM tunnels built on soft soil in seismic areas, while the TMG and TMF concepts, for TBM railway and road tunnels, respectively, adequate arrangements of the tunnel's (circular) cross section and the adoption of additional innovative measures, allow a significant improvement in safety during operation and at the same time a significant reduction in construction costs.

The main objectives of the book are, on the one hand, to present the current situation of tunnel construction in the world, in particular using the TBM technique, and on the other, to present the concepts mentioned above and show the benefits that can be obtained with their application in the construction of TBM traffic tunnels. It also intends to drawn attention to other innovations in TBM tunneling that may improve the construction of tunnels in the future.

Chapter 1 presents the "world of tunneling" in historical perspective, referring to the most significant events in tunnel construction throughout the History and the most common tunneling techniques used today.

Chapter 2 presents the most significant aspects related to the construction of TBM tunnels, namely the development of the machines and the types of existing TBMs, according to the characteristics of the ground to be bored, referring also to some other specific issues.

Chapter 3 presents the main safety requirements of rail and road tunnels, as reflected in the most significant existing normative documents, mainly at EU level, as well as an analysis of the most significant issues related to compliance with these safety requirements.

Chapter 4 presents the most advanced current concepts for TBM traffic tunnels, as well as noteworthy examples of railway and road tunnels built in recent times based on those concepts, focusing on aspects such as geometry, ventilation, fire fighting and rescue of users in the event of an accident or fire, mentioning the main achievements and shortcomings.

Chapter 5 is dedicated to the presentation of the TISB, TMG and TMF concepts for TBM tunnels, as well as the most relevant technical issues related to the conceptual design and construction of tunnels based on these concepts.

Chapters 6 to 9 present examples of the application of the TISB, TMG and TMF concepts to the construction of TBM traffic tunnels, respectively: application of the TISB concept in an alternative solution for the rehabilitation of a TBM metro tunnel; application of the TMG concept to improved TBM tunnel solutions for HSRL (High-Speed Railway Lines); application of the TISB and TMF concepts in an improved solution of a TBM road tunnel on soft ground in a seismic area; and application of the TMG and TMF concepts in an alternative TBM tunnel solution for a large mixed rail and road crossing, showing in each case the benefits that can be obtained by applying these concepts.

Chapter 10 presents a summary of the main conclusions drawn in the previous chapters, as well as the perspectives that open up with the application of the TISB, TMG and TMF concepts in new challenges, mentioning also other technological advances that may improve the construction of TBM traffic tunnels in the future: robotization of risky operations, TBM docking, application of BIM tools, etc., especially when the above mentioned concepts are used.

Chapter 1

The world of tunnelling

Abstract

This chapter provides an overview of tunnel construction in the world throughout History.

A summarized history of tunnel construction is presented, referring to the main events over the different historical periods, as well as the main purposes of the tunnels.

The different tunnel construction techniques are described: cut-and-cover tunnels, sequential excavation tunnels, immersed tunnels and TBM tunnels, referring to their specificities.

Landmarks in tunnel construction worldwide are also presented.

1.1 INTRODUCTION

Tunnels are underground passages, dug through the surrounding soil, earth or rock, usually closed, except at the entrance and exit, commonly placed at each end.

Tunnels, normally serve as an infrastructure in mobility systems, for pedestrian traffic, for road vehicular traffic, for rail traffic (common rail or metro systems), or for a canal. They can also serve as aqueducts to supply water for consumption or for hydroelectric plants, as sewers, or to house telecommunication networks.

The importance of tunnels is great today but it will probably be even greater in the future, playing a decisive role in building a more "green" and sustainable world, facilitating traffic and reducing pollution in urban areas.

1.2 BRIEF HISTORY OF TUNNELLING

The history of tunnelling is presumed to be almost as old as Humanity itself, with testimonies of tunnel building dating back at least 5,000 years.

In tunnelling three main periods can be considered: the Ancient Era, from Antiquity to the last quarter of the 18th century; the Industrial Era, from that time until the 1950's; and the Modern Era, from the 1950's to the present.

As one might expect, at the beginning, most tunnels were built for use as tombs and/or religious practices where builders were free to choose their locations, so as to avoid difficult-to-excavate ground or the need for ground support. Also, natural and excavated caves were used as shelter, in which there are still traces of this use today.

There are also old traces of the excavation of tunnels and wells for mineral extraction, as in the case of the "Grimes Grave" mine, in England, during the Mesolithic period, about 1500 BC, in which deer horns were used as excavation tools.

DOI 10.1201/9781003144861-2

Over time, tunnels have also started to be used for drainage, water supply, sewage and transport projects (including canals) where the location of the tunnel has become fundamental and various, sometimes difficult, ground conditions have to be managed.

The oldest tunnel recorded for use as such was built between 2180 and 2160 BC in Babylon, under the Euphrates River. The river was diverted from its original bed and the technique known today as "cut-and-cover" was used. The purpose of this tunnel was to connect the royal palace to the temple, which was about 1 km away.

The Egyptians also paid great attention to the construction of tunnels, developing special techniques for cuting soft stones for use in their temples and to built tombs, as in the Valley of the Kings, in Luxor, from the 16th to 11th century BC.

The Grecs also made extensive use of tunnels. The Eupalinian Aqueduct on the island of Samos (Eastern Aegean, Greece), about 1.0 km long, was built in 520 BC by the Greek engineer Eupalinos of Megara, regarded as a major feat of ancient engineering (figure 1.1) (Wikipedia 2021a).

Figure 1.1.   The Eupalinian Aqueduct in Samos island, Greece

Eupalinos, knowledgeable about geometry, organised the work so that the tunnel started on both sides of Mount Kastro and the two teams advanced simultaneously and met in the middle, with excellent precision, something which would not have been easy at that time.

The aqueduct was of extreme defensive importance, as it ran underground, and would not be easily found by an enemy who wanted to cut-off the water supply to Pythagoreion, the ancient capital of Samos.

The existence of the tunnel was recorded by Herodotus (as well as the mole and harbour, and the great temple to Hera). Rediscovered in the 19th century, visitors were able to enter it from the 1970's.

One of the first known drainage and sewer networks in form of tunnels, was built in Persepolis, Iran, at the time of its foundation in 518 BC. In most places the network was dug in the sound rock of the mountain and covered with large pieces of rock and stone, followed by earth and piles of rubble to level the ground.

The oldest tunnel in the world built under a body of water is claimed to be the Terelek kaya Tunnel on the Kizil River, south of the cities of Boyabat and Durağan in eastern Turkey, more than two thousand years ago.

Currently, the tunnel is under a narrow part of a lake formed by a dam a few kilometers downstream. It is assumed that the tunnel was built for a defensive purpose by the same people who built the royal tombs on a nearby rock face.

During the Roman Empire, between 95 and 326 AC, a large network of tunnels was built. The Via Flaminia, an important Roman road, penetrated the Furlo pass in the Apennines through a tunnel that the emperor Vespasian had ordered to build between 76 and 77 AC. A modern road, the SS3 Flaminia, still uses this tunnel, which had a precursor that dates back to the 3rd century BC; remnants of that previous tunnel (one of the first known road tunnels) are also visible.

Also noteworthy are the approximately 940 km of underground galleries in Rome, connecting the 60 catacombs, which housed millions of Christians during the persecutions they were subjected to during the 2nd century, until the middle of the 4th century AC (figure 1.2) (Wikipedia 2021b).

Figure 1.2.   Tunnels connecting the catacombs under Rome

During the Middle Ages the construction of tunnels declines, being built essentially for military purposes, whether offensive or defensive, to pass under the walls that defend the cities, or to escape the assault of enemies.

In the last quarter of the 18th century, with the (1st) Industrial Revolution and the appearance of steam engines in 1777, a new era began in the tunnel construction: the Industrial Era, one of the most productive times for tunnel engineering, with the need to pave the way for trains to pass.

In the Industrial Era, however, two periods have to be considered; the first, corresponding to the first century of the Industrial Revolution (until the 1870s), where traditional hand-mined techniques were used, and the second, from that time until the 1950's, where several important innovations were introduced in tunnelling technology.

As landmarks of the construction of tunnels at that time, the following are mentioned.

Once the Industrial Revolution began in England, it is not surprising that most of the first major tunnels were built there. It was normal the use of black powder and various wooden methods. The final linings were made with bricks or dressing stone and cement, but only when absolutely required by bad ground. The vast majority of tunnels were built in rock, the first of which were still integrated into canals for river traffic.

Thus, in 1789 the Sapperton Canal Tunnel was opened on the Thames River, 3.5 km long and used to transport coal and other goods in small boats.

However, although relatively short, only 165 meters long, the world's first tunnel canal in the world was built in France, a century earlier, in 1769, the Malpas Tunnel (figure 1.3) (Wikipedia 2021c), on the Canal du Midi, classified as World Heritage in 1996.

Figure 1.3.   The Malpas Tunnel, on the Canal du Midi, France

Also in England, in 1793 it was built the Fritchley Tunnel in Derbyshire, to transport limestone wagons using gravity and horse traction. In 1813 the railway was converted to steam locomotion, the tunnel being used continuously untill 1933, when the line was closed. During World War II, it was used as bomb shelter. In 1977 it was classified as an Ancient Monument and sealed to preserve the construction.

The 1796 Stoddart Tunnel in Chapel-en-le-Frith in Derbyshire, also in England, is reputed to be the oldest railway tunnel in the world, although the railway wagons were also originally also drawn by horses.

Across the Atlantic, in the USA, the famous "Derby Tunnels" in Salem, Massachusetts, were built in 1801 for a peculiar purpose: to smuggle imports affected by President Thomas Jefferson's new customs duties. Jefferson ordered local militias to assist Customs at each port to collect these dues, but smugglers hired the militias to dig tunnels and hide spoils. The tunnels covered 4.8 km connecting the city's wharves to an underground train station. Along the way, they connected homes of prominent businessmen and politicians to their stores and banks.

At that time, a tunnel was built also in England for the first real steam locomotive, from Penydarren to Abercynon, which made the historic journey in 1804. This is the oldest railway tunnel in the world, dedicated only to self-propelled steam engines on rails.

During the 1820's and 1830's several tunnels were built in England, to connect railway lines in several busy cities, as well as to connect the cities of Liverpool and Manchester.

The Box Tunnel, on the Great Western Main Line, between Bath and Chippenham, opened in 1841, 2.9 km long, was considered the longest railway tunnel in the world at the time of its construction.

In 1842 opened the Victoria Tunnel in Newcastle, also in England, a 3.9 km long underground wagonway with a maximum depth of 26m, which descends 68m from entrance to exit. It runs under Newcastle upon Tyne, being originally designed to carry coal from Spital Tongues to the river. During the World War II, part of the tunnel was used as a shelter.

In 1843, after 18 years of works, the Thames Tunnel was opened in London, under the Thames River, which was the first tunnel (after the Terelek Tunnel) built traversing a body of water. It runs 23 meters below the river surface, being built by the French engineer Marc Isambard Brunel, the first time that a tunnelling shield was used (figure 1.4) (Wikipedia, 2021d), a concept that he patented in 1818.

Figure 1.4. The tunneling shield patented in 1818 by Marc Brunel

The tunnel was originally used as a pedestrianway (figure 1.5) (Wikipedia 2021d), being converted to railway traffic in 1869, as part of the East London Line of the London Underground until 2007. It was the oldest section of the network, although it was not the oldest built for that purpose.

Despite the difficulties faced in its construction, the tunnel turned out to be a technical success, showing that it was possible to build underwater tunnels, being at the time described as the "eighth wonder of the world" and further recognised as an "International Historic Civil Engineering Landmark".

Figure 1.5.   The Thames Tunnel, in the middle of the 19th century

However, despite the improvements that have been made, the shield concept for tunnel building was not convincing for a long time, until the middle of the 20th century, in 1952, when the American engineer James S. Robbins associated the shield with the rotary cutting head, creating the first TBM ("Tunnel Boring Machine").

In 1848 was the opening of the Victoria/Waterloo Tunnel, in Liverpool, England, 3.34 km long. The tunnel was initially used only for rail freight, serving the Waterloo Freight terminal and later freight and passengers serving the Liverpool ship liner terminal.

Also in 1848 it was opened the first alpine tunnel, the vertex tunnel of the Semmering railway in Austria, 1.43 km long, creating a railway link between Vienna, capital of Austro-Hungarian Empire, and Trieste, its port. In 1998 it was classified World Heritage site.

Shortly thereafter, in 1854, in what would become Italy, opened the Giovi Railway Tunnel, accross the Apennine Mountains, 3.25 km long, connecting Turim, the capital of the then Kingdom of Sardinia, to its port, Genoa.

In 1868 it was opened the 0.5 km long Summit Tunnel, at the Donner Pass, in the Sierra Nevada Mountains, in California, USA, allowing the establishment of mass commercial transport of passengers and freight to Sacramento, in the Pacific coast. It remained in daily use until 1993, when all the rail traffic in the area was transferred to the 3.15 km long "The Big Hole" Tunnel, 3.1 km long, built in 1925.

In 1870, after 14 years of works, the railway Fréjus Tunnel (also known as the Mont Cenis Tunnel), between France and Italy, was completed, a dual-track single-tube tunnel, the second oldest alpine tunnel and the first long railway tunnel in the Alps, 12.8 km long, the longest in the world at the time.

At the beginning of its construction, the first known tunnel boring machine as such was used, called "Mountain Slicer", developed by the Belgian engineer Henri-Joseph Maus in 1845, however the tunnel ended up being built mostly by conventional means.

By that time, other important innovations were introduced in the tunnelling activity, which represent the begining of a new period of progress in tunnel construction.

In 1878, the shield technology took a big step forward with a provisional patent developed by a certain Mr. Barlow, in which four main features were adopted: a cylindrical overlapping skin, a final cast iron lining, the placement of grout outside the lining, and the use of screw jacks to move the entire shield forward as a single unit.

Other important innovations also made their debut at that time, such as the pneumatic rock-drilling machine, patented by Germain Sommeiller (figure 1.6) (Wikipedia 2021), dynamite, cladding construction and various subsidiary developments, such as ventilation and air control survey, which allowed the construction of large tunnels through the Alps. In addition, the shield technology combined with compressed air was showing be very promising for excavation and support in all types of soil conditions.

Figure 1.6.   The pneumatic rock-drilling machine used to bore large tunnels through the Alps

Thus, in 1882, the third alpine tunnel, the Gotthard Railway Tunnel, was opened connecting the north and the south of Switzerland, 15 km long, the new longest railway tunnel in the world.

Still in 1882, it was also inaugurated the Col-de-Tende Road Tunnel, running between France and Italy, 3.2 km long, the first long road tunnel under a pass.

In January 1886, the railway Mersey Tunnel, which runs from Liverpool to Birkenhead, was inaugurated, 5 km long, of which 1.2 km under the River Mersey, being the longest underwater tunnel in the world at the time.

In late 1886, the railway Severn Tunnel, 7.0 km long, of which 3.6 km are under the River Severn was opened, replacing the Mersey tunnel as the world's longest underwater tunnel.

In 1890, the City & South London Railway tunnel was opened under the Thames River, built by James Greathead, which brought together three key elements for the construction of underwater tunnels: 1) shield method of excavation; 2) permanent cast iron tunnel lining; and 3) construction in a compressed air environment to inhibit the flow of water through soft ground material into the tunnel portal.

In 1900, in France, after several hesitations and discussions, the first lines of the Paris metro were opened, built using the cut-and-cover method.

In 1906, the fourth alpine tunnel, the Simplon Tunnel, between Switzerland and Italy, was opened, with electric traction rather than steam from the beginning.

With two separated tubes (one for each track) and 19.8 km long it was the longest tunnel in the world for a long time. It was also the deepest tunnel in the world, with a rock cover of 2150 meters. However, the second tube came into operation only in 1922.

In 1927, it was inaugurated the roadway Holland Tunnel (figure 1.7) (Wikipedia 2021e), under the Hudson River, in New York City, USA, formed by two parallel galleries, 2.6 km long each.

Figure 1.7.   Holland Tunnel. Construction in progress, 1923

It was the first underwater tunnel designed for road traffic, requiring a new ventilation system, with ventilation towers on each bank, being the first mechanically ventilated tunnel in the world.

In 1945, the Delaware Aqueduct tunnel was completed, starting to supply water to New York City, USA. At 137 km long, it is still the longest tunnel in the world.

In 1952, as mentioned, James S. Robbins revolutionized the construction of tunnels, associating the shield to the rotary cutter head, creating the first modern TBM (figure 1.8) (Robbins 2021), which successfully excavated in weak shale rock at the Oahe Dam Project.

In 1956 Robbins went even further and invented the rolling disc cutter, which managed to drill the first rock tunnel ever, in the Humber River project in Canada.

These achievements represented a major step towards the mechanisation of tunnel construction, inaugurating the so-called Modern (Tunnelling) Era.

At that time, other important developments occurred, namely the emergence of Engineering Geology, and the subsurface exploration, largely based on the work of Karl Terzaghi, recognized as the founder of Soil Mechanics.

Terzaghi recommended intense geological mapping exercises and subsurface exploration programs and, during construction, field observations in order to make sure that the ground was behaving as anticipated.

Figure 1.8. The Robbins 1952, the first modern TBM

In terms of technologies, the Modern Era in tunnelling is characterized by the widespread use of the tunnel boring machines (TBMs) but also advances in other techniques, from conventional ones to the immersed tunnels, shotcrete and soil improvements, which had profound impact in reducing costs, risks and schedules in the construction projects. In Section 1.3 the most significant current tunnelling techniques are presented and commented on.

Regarding TBMs, the development in the 1960's and 1970's of the Pressurized Face Shield machines (the Slurry Shields and the Earth Pressure Balance) greatly enlarged the types of soil that could be excavated safely. Detailed information about the methods currently used by the TBM technique is given in Chapter 2.

Some highlights have to be mentioned.

In 1965, the mono-tube bi-directional roadway Mont Blanc Tunnel under the Alps was opened to traffic, between France and Italy, 11.6 km long, three times longer than any existing road tunnel at the time. It was excavated by conventional methods (drill-and-blast, etc.), 2430 meters below the surface, the deepest tunnel at the time and still currently the second deepest, after the Gotthard-Base Tunnel.

In 1988 was completed the mono-tube double-track railway Seikan Tunnel, under the Tsugaru Strait, in northern Japan, 53.8 km long, connecting the Honshu and Hokkaido islands, becoming the longest railway tunnel in the world, also built using conventional tunnelling methods. It is provided with a service tunnel along the under-water section and emergency stop stations on both shores.

In 1994, was opened the three-tube single track railway Channel Tunnel, between France and UK, under the English Channel (see, Section 4.2.2), 50.5 km long, from which 37.9 km under water, still the longest underwater tunnel in the world, built with the TBM technique, in whose construction 11 TBMs were used.

In 2000 was opened to traffic the mono-tube dual-lane Laerdal Tunnel, in Norway, 24.5 km long, the longest road tunnel in the world, which was built by conventional methods.

In 2013 a shielded TBM with an excavation diameter of 15.60 metres, the largest TBM in hard rock, completed in 24 months the two tubes of the roadway Sparvo Tunnel, 5 km long, with three lanes in each direction, in the Appenins, near Florence, Italy (figure 1.9) (Revetec 2021).

Figure 1.9.   Interior view of one of the tubes of the Sparvo Tunnel

In 2016, the dual-tube twin-track railway Gotthard-Base Tunnel was opened to traffic in Switzerland, under the Alps (see, Section 4.2.3), with 57 km in length and 2450 meters of rock cover, built using mainly TBMs, the longest and deepest traffic tunnel in the world.

Also in 2016, the roadway mono-tube Eurasia Tunnel was opened to traffic under the Bosphorus Strait, in Instanbul, Turquey (see, Section 4.3.3), going 106 meters below sea level, built with a Mixshield TBM, prepared for pressures up to 12 bar, the most in the world.

In 2017, the dual-tube road Ryfast Tunnel, near Stavanger, Norway, with a length of 14.3 km, excavated by conventional methods, became the longest underwater road tunnel in the world, surpassing the tunnel under the Tokyo Bay, Japan, 9.6 km long, and the Shangjiang Tunnel in Shanghai, China, 8.95 km long (see, Chapter 4).

In 2018, it opened to traffic the Hong Kong-Zhuhai-Macau Roadway Fixed Link, in China, 55 km long, the world's longest sea crossing, which includes an immersed tunnel, 6.7 km long, the longest immersed tunnel in the world.

1.3   PURPOSE OF TUNNELS

As mentioned, tunnels are mainly for pedrestrians, for road or rail traffic, or for a canal. They can also fuction as aqueducts to supply water for consumption or for hydroelectric plants, or as sewers. Tunnels for rail traffic can be for urban or rapid transit, or for main-line railways.

Some traffic tunnels house mixed traffic (road and rail), with superimposed galleries, the upper one for road traffic and the lower one for a railway or a subway line. Traffic galleries can also be placed side by side. However, mixing different traffics in the same tunnel present risks, as it can lead to conflicts in the respective safety concepts.

There are also the so-called utility tunnels, used for routing steam, cold water, electrical power or telecommunication cables, as well as connecting buildings for the passage of people and equipment.

Sometimes utility tunnels are MUTs ("Multi-Purpose Utility Tunnels"), that is utility tunnels which carry two or more utility lines. By placing different utilities in the same tunnel, it will be possible to reduce the construction and maintaining costs of those utilities.

There still are tunnels, such as wildlife crossings, built to allow wild animals to cross human-made barriers, safely.

There are tunnels with non-conventional purposes, like the secret tunnels, built for military purposes, or by civilians for shelter or for smuggling of products or people.

There are also special multi-purpose tunnels, such as the SMART ("Stormwater Management and Road Tunnel") in Kuala Lumpur, Malaysia, 4.0 km long and three superimposed galleries, to convey road traffic in both directions and drain occasional flood water. In a first stage, the flood water is diverted to the gallery located at the tunnel base, while road traffic continues normally in both road galleries. During heavy and prolonged rains, when the threat of extreme flooding is high, the traffic galleries are closed to vehicles and automated flood control gates are opened so that water can be diverted through both road galleries.

1.4 TUNNEL CONSTRUCTION TECHNIQUES

1.4.1 General

There are available several tunnelling techniques as well as variants, which refer to the specialization of a particular technique, according to the type of ground to be excavated.

The evolution of these techniques results from a practical knowledge obtained over the years and the need to build increasingly longer tunnels, in geologically more complex massifs, with higher safety and more economically viable, which was shaping and, therefore, making these techniques progressing.

Tunnels are today dug in all types of terrain, from soft clay to hard rock. The appropriate method to be used in the construction of a tunnel depends on several factors, such as ground conditions, groundwater conditions, the length and diameter of the tunnel, the depth of the tunnel, the support logistics of the tunnel excavation, the final use and the shape of the tunnel, and appropriate risk management.

The better understanding of the massif's behaviour, which has been obtained through studies that can be carried out with modern technology, has raised these techniques to the level known today. In the following it will be presented and briefly explained the different tunnel construction techniques currently in use.

Nowadays, basically four tunnel construction techniques are used: cut-and-cover tunnels, sequential excavation tunnels, immersed tunnels and bored (TBM) tunnels, each of them with their specific conditions of use. The most relevant aspects of these techniques are presented below.

Also worth mentioning is the freezing tecnique, sometimes used in the construction of small galleries (cross-passages) connecting the main tubes of multi-tube underwater tunnels, which will be mentioned in Chapter 4, when presenting noteworthy examples of tunnels.

Shallow tunnels are generally of the cut-and-cover type or of the immersed type (when under water), while deep tunnels are usually excavated using mining techniques and increasingly bored, using shielded TBMs.

1.4.2   Cut-and-cover tunnels

With the cut-and-cover technique the structure of the tunnel is built inside the excavated area and later covered with material extracted from the excavation. This technique is usually used when the tunnel is relatively shallow (6 to 12 meters deep).

It is, however, to be note that this technique implies a great interference with the surface, that is, it is not possible to build a tunnel with this technique without altering the surface, which in some cases, particularly in urban spaces, affects in large scale the normal functioning of these areas.

This excavation technique comprises two different methods, in terms of procedures: Bottom-up ("From the bottom to the top") and Top-down ("From the top to the bottom"), each of which has advantages and disadvantages.

Bottom-up method

In this method, a trench is dug along the alignment in which the tunnel will be built, inside which the tunnel is subsequently built and its construction completed. The trench is then carefully back-filled and the situation on the surface is reinstated.

This trench can be made by digging in depth, making slopes without support, when there is space on the surface that allows it. In the case of lack of space on the surface or when there is a need to control the displacement of the ground, a peripheral containment is carried out, which may consist of pile curtains, molded walls, sheet piles or even Berlin-type containment.

Figure 1.10 (FHWA 2009) presents the general excavation sequence according to this method:

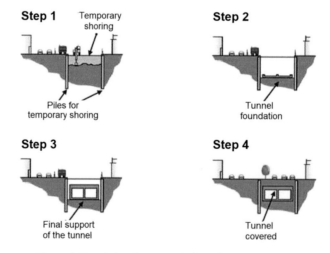

Figure 1.10.   Cut-and-cover technique. Botom-up method

Step 1: Installation of temporary excavation support such as molded piles, lowering the water table (if necessary), and excavation and installation of support for temporary walls, for example struts;

Step 2: Construction of the tunnel foundation and (if necessary) improvement of the foundation mass;

Step 3: Finalization of the construction of the tunnel body and installation (if necessary) of waterproofing elements;

Step 4: Filling the stretch and restoring the surface.

This method has several advantages, such as being easily mastered by tunnel builders, possibility of applying waterproofing elements outside the tunnel, possibility of installing drainage and/or water diversion systems outside the tunnel and easier access to the interior of the excavation compared to the Top-down method.

As disadvantages, refer to the following: it requires more surface area to carry out the excavation, impossibility of restoring the surface until the completition of the construction, requires provisional support and the need to lower the water table, which may cause adverse effects on the environment.

Top-down method

In this method, also known as "inverted method", the tunnel walls are built first, using, for example, drying molded piles, that is, the excavation support is the final support of the tunnel.

Then, the top slab of the tunnel is built, supported by the previously executed walls, and the use of the surface is restored.

The rest of the excavation, construction of the tunnel and finishes are carried out under the top slab, thus allowing the early return to surface regularity.

In figure 1.11 (FHWA 2009) the general excavation sequence of this method is presented:

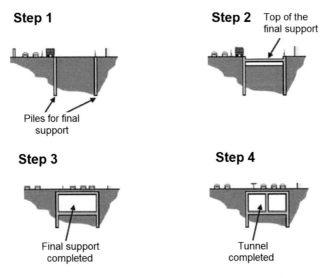

Figure 1.11.   Cut-and-cover technique. Top-down method

Step 1: Installation of the excavation final support, for example drying molded piles, lowering the water table (if necessary);

Step 2: Excavation of the top of the stretch, construction of the slab at the top of the tunnel and (if necessary) installation of waterproofing elements over the slab;

Step 3: Filling the excavated section and restoring the surface, excavation of the interior of the tunnel and anchoring of the tunnel walls (if necessary), execution of the tunnel foundation and improvement of the foundation (if necessary);

Step 4: Execution of finishes inside the tunnel.

This method as advantages: it allows the early restoration of the surface; the provisional support is normally used as the final support for the tunnel; the structural slabs act as struts allowing the reduction of anchors in the tunnel walls; requires less space on the surface than the Bottom-up method; it allows to reduce the construction time, since it allows the execution of several tasks simultaneously; it can result in a lower overall cost due to the possible decrease in the support to be used; and enables reduction in construction time.

The disadvantages of this method are as follows: impossibility of installing waterproofing elements outside except on the top slab of the tunnel; possibility of infiltrations in the joints of structural elements; the connections between slabs and walls are more difficult to perform; risk of deviation of the walls into the tunnel during the construction of the walls; limited access to the excavation and limited space for excavation on the interior of the tunnel and the construction of the lower slabs.

1.4.3 Sequential excavation tunnels

The so-called Sequential Excavation Method (SEM), also known as the New Austrian Tunnelling Method (NATM), is based on the knowledge of the massif's behaviour when creating an opening in it, being an advance in the conventional mining techniques of tunnel construction.

The main idea of this technique is to take advantage of the massif's self-support capacity, in order to decrease the amount of immediate support. After each section is excavated using conventional mining techniques, it is protected by a layer of sprayed concrete, commonly called shotcrete. Other supportive measures may be added, such as steel arches, rock bolts, and wire mesh. Sometimes, steel or polypropylene fibers are also added to the concrete mix to improve the strength of the lining.

After opening the entire tunnel, the final support is then placed in order to satisfy the purpose of the tunnel.

With adequate monitoring, the method is flexible, even in the face of surprising changes in the geomechanical consistency of the rock during the tunnelling work. The measured properties of the rock lead to the choice of appropriate tools for strengthening the tunnel.

This technique, initially developed in the 1960s to excavate in hard rock, has also started to be used in earth massifs in urban areas. It is currently used in both types of massifs, in urban or rural environments. The method's ability to adapt to various geometries while maintaining safety and economy levels has made it widely used.

The phasing of the tunnel excavation depends on several factors, mainly the geological, geotechnical and hydrogeological conditions of the massif, the cross-section of the tunnel and the possibility of the presence of water. Then, it is assumed that the greater the cross-section of the tunnel and the lower the geomechanical capacity of the massif, the more numerous the excavation phases should be. Figure 1.12 shows two different partial excavation sequences, using the NATM method.

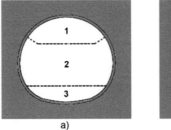 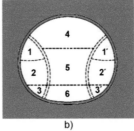

a)   b)

Figure 1.12.   Examples of excavation sequences with the NATM method

When the massif to be excavated is earthy, without high stresses in situ, or rocky in which the rock is relatively soft, little abrasive and with some fracture, it is possible to use punctual attack excavation techniques, such as brushcutters and excavators equipped with buckets or pneumatic hammers.

If the massif is formed of high resistance rock, the use of explosives will be necessary. This use implies the need to carry out feasibility studies, in order to predict and prevent possible damage that may occur to structures existing on the surface or underground in the surrounding area, as well as to the massif itself.

The execution of a disassembly with explosives is essentially related to the application of a fire diagram. The choice of the correct explosive for a given mass is a challenging issue, since, in addition to technical matters, economic, environmental and safety matters must be taken into account.

1.4.4   Immersed tunnels

The Immersed tube tunnels, also called Sunken tubes, or simply, Immersed tunnels, are a tunnelling technique used mainly for underwater crossings in shallow water.

This method was pioneered by the American engineer W. J. Wilgus, on the Detroit River, in 1903, for the Michigan Central Railroad, and since then, about a hundred immersed tunnels have been built around the world (De Wit 2014).

In this method, the tunnel is made up of several elements of great length, which are built on land and then sunk in a pre-dredged trench in the seabed of the waterway, where they are connected to each others at the ends.

The conventional method of constructing an immersed tunnel is to establish one or more casting basins, where the individual tunnel elements are built, being usually composed of several segments, which are connected to each others.

When the elements are completed, they are temporarily sealed at the ends with bulkheads placed in the casting basins and afterwards flooded one by one in the intended location, immersed into their final position on the bottom of the trench and connected to those already installed.

The connection to the land surface is normally made through cut-and-cover sections that bridge the gap between the water bottom and the land surface.

Immersed tunnels mainly serve as rail or road tunnels, being also sometimes used for water supply and electric cables (Foley 2013).

Tubes are usually rectangular, but they can also be circular or oval. In large crossings, wide rectangular shapes are usually considered to be more cost-effective (figure 1.13) (Foley 2013 (adapted).

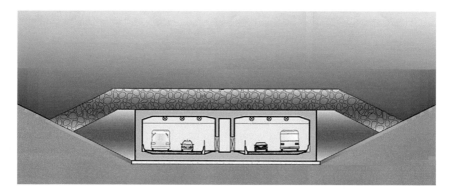

Figure 1.13.   Typical cross-section of an Immersed tunnel for road traffic

Tunnel construction operations are thus distributed between the casting basin(s) and the immersion of the tunnel elements on the site.

In the casting basin(s): construction of the casting basin; construction of the first batch of segments; construction of the tunnel element connecting the segments; construction of the temporary water ballast system; sealing the element with a temporary bulkhead; filling of the casting basin (flood); removal of the basin gate, float and winch of the element out of the basin and transport to the immersion site, one by one, or to a temporary parking place.

For the immersion process: dredge the trench to install the tunnel elements; build a bed of gravel at the bottom of the trench; immerse the tunnel element in its specific place in the dredged trench; connection of the element by pumping and emptying the water from of space between the bulkheads (water pressure on the free end of the new element compresses and seals the rubber seal between the two elements); placement of backfill material on the sides and over the tunnel; fill the trench and permanently bury the tunnel; ballast the element on-site; and, installation of the final closing joint. In parallel with the immersion process, the tunnel is being interiorly finished on the elements already installed.

The main advantage of immersed tunnels is that, in some situations (mainy in short tunnels) they can be more cost-effective than alternative options, for example, a TBM tunnel or a bridge, as these need to be more lengthy. Other advantages include: the speed of construction, the safety of construction (for example, work on a dry dock as opposed to boring under water) and the flexibility of the cross-section, although this is usually dictated by the conditions at the ends.

The disadvantages of immersed tunnels include: they are often partly exposed on the river/sea bed, risking a ship/anchor strike; careful waterproofing design of the joints; careful design of the connections for the longitudinal and transverse forces that have to be transferred across; disruption to the river/channel traffic during the immersion process; often, the need for the strengthening of the foundation soil; and, mainly, the environmental impact on the underwater embankment and existing channel/sea bed.

1.4.5 Bored (TBM) tunnels

TBM tunnels, that is, tunnels built by a TBM ("Tunnel Boring Machine") and its associated backup system, seek to highly automate the entire tunnelling process, allowing to faster construction and lower tunnelling costs.

The concept of the operation of a tunnel boring machine comes, as mentioned above, from the combination of the tunnelling shield developed by Marc Isambard Brunel in 1828 and the rotary cutting head created by James S. Robbins in 1952.

The TBM technique has gained increased importance, because, beyond the economic advantages, its use do not conflict with the existences (very low environmental impact), and because allows for the possibility of crossing complex geological and hydrological conditions safely and with relative economy (figure 1.14) (Wikipedia 2021f).

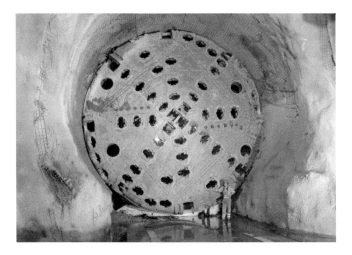

Figure 1.14.   Front-shield of one of the TBMs used on the Gotthard-Base Tunnel, in Switzerland

In addition, in predominantly urban applications, the TBM technique is a quick and economical solution for the creation of rail and road connections, since it eliminates interference with surface activities, and thus the eventual need for land acquisition, as well as with consultations potentially long planning periods.

The disadvantages of TBMs arise from their usually high cost and large size: the difficulty of transporting the large TBM components to the tunnel construction site and of assembling them on-site, often in very limited space.

Despite these difficulties, TBMs are now the generally preferred method, especially since, with technological progress, the old compressed air tunneling method with an airlock/decompression chamber in the front of the TBM, which required operators to work in high pressure and undergo decompression procedures at the end of their shifts, were overpassed.

Some types of TBMs, such as the earth pressure balances and the bentonite slurry machines, have pressurized compartments in the front, allowing them to be used in difficult conditions below the water table, pressurizing the ground ahead of the TBM cutter head to balance the water pressure.

Thus, the operators work with normal air pressure behind the pressurized compartment, only occasionally entering that compartment to renew or repair the cutter devices.

There are currently several types of TBMs, according to the ground to be excavated, that can operate in virtually all types of conditions, from hard rock to soft soils with water.

Chapter 2 provides more detailed information on the different types of TBMs.

1.5   LANDMARKS IN TUNNEL CONSTRUCTION

With regard to the tunnel length, the following cases are worth mentioning (Wikipedia 2021g).

The longest tunnels in the world are for water supply.

The Delaware Aqueduct tunnel, which has supplied water to New York City, USA, in service since 1945, is the longest, 137 km long. The Päijänne Water Tunnel, in southern Finland, opened in 1982, is the second longer, 120 km long, and the Dahuofang Water Tunnel, in northeaster coastal China, opened in 2009, is the third longest, with 85.3 km.

The longest traffic tunnels are railway tunnels.

The Gotthard-Base Tunnel, in southern Switzerland (see, Chapter 4), is the longest and deepest tunnel in the world, 57.1 km long and 2450 meters maximum depth below the Saint Gotthard massif.

The Seikan Tunnel in Japan, which connects the main island of Honshu to the northern island of Hokkaido, is the second, 53.9 km long, of which 23.3 km under the sea, crossing the Tsugaru Strait.

The Channel Tunnel, which crosses the English Channel, between France and England (UK) (see, Chapter 4), is the third longest, with 50 km in length, of which 39 km under the sea, thus being the largest underwater tunnel in the world.

Regarding road tunnels, the Lærdal Tunnel, in Norway, between Lærdal and Aurland, is the longest in the world, 24.5 km in length. The Zhongnanshan Tunnel in China, with 18 km in length, is the second longest and the longest highway tunnel in the world.

In Chapter 4 more details about some of these tunnels are presented.

1.6   CONCLUSIONS

It can be said that the Tunnelling history is confused with the history of mankind and its progress and development over the centuries.

The tunnels were first built as human shelter and later for utilitarian purposes, mainly to transporting water for human consumption or for irrigation the fields.

With the Roman civilization tunnels began to be built to improve the traffic connections, integrated in the Roman road network.

During the Middle Ages the interest in the construction of tunnels decrease, which start to be built essentially with a defensive purpose, to pass under the walls that defend the cities.

At the last quarter of the 18th century, with the start of the Industrial Revolution, there was a great increase in the construction of tunnels, flowing the great development of traffic, mainly with the emergence of the steam engine and the railway networks.

In the early 1800s, the first tunnels for railway traffic were built.

For a long time, tunnels were built essentially on rocks (hard or soft), using the conventional mining techniques.

In the mid-19th century, the invention of the tunnel shield by Marc I. Brunel, patented in 1818, the first underwater tunnel, the Thames Tunnel, under the Thames River in London, was built, opening the possibility of building tunnels in difficult ground conditions.

However, despite the recognition of this great achievement, the tunnel shield was not been completely convincing at the time and it was necessary to wait until the middle of the 20th century, when James S. Robbins, in 1952, revolutionized the construction of tunnels, creating the modern TBM ("Tunnel Boring Machine") concept, mechanizing practically all necessary operations.

In the meantime, by the end of the 19th century, other innovations were introduced in tunnelling, allowing significant improvements in the construction of tunnels, such as dynamite and compressed-air drills, as well as the debut of the immersed tunnel technique.

Currently, different tunnel construction techniques are in use, each one suited to the specific conditions of the terrain and location, basically the cut-and-cover technique, the sequential excavation technique, the immersed technique and the TBM technique.

As regard the TBM technology, there was an authentic revolution during the last decades, now having machines adapted to build tunnels in all types of conditions, safely and economically, in particular traffic tunnels, allowing the construction of increasingly larger tunnels in short periods of time.

REFERENCES

Benvenuti G (2015), *Un Tratto della Variante di Valico dell'Autostrada A1, Inaugurata Oggi (A Section of the A1 Motorway Valico Variant, Inaugurated Today)*. ANSA (in italian), 23 December 2015

De Wit J C; Van Putten E (2014), *The Immersed Tunnel as Fixed Link. A Successful Alternative Pushed by Innovation*. World Tunnel Congress (WTC2014) Tunnels for a Better Life, Foz do Iguassu, Brazil, May 2014

FHWA (2009), *Technical Manual for Design and Construction of Road Tunnels - Civil Elements*. U. S. Department of Transportation, Federal Highway Administration, Publication FHWA-NHI-10-034, December 2009

Foley A (2013), *Scoops Brazil's First Immersed Tunnel*. Tunneling Journal, April 2013

Gary Brierley (2014), *Tunneling: A Historical Perspective*. Tunnel Business Magazine, August 2014

Green A (2014), *Just Keep Digging: A Brief History of Tunnels*. Popular Mechanics, January 2014

He C, Wang B (2013), *Research, Progress and Development Trends of Highway Tunnels in China*. Journal of Modern Transportation, Volume 21, 2013

Lane K S (2021), *Tunnels and Underground Excavations*. Encyclopedia Britannica, 2021

Nôro A M (2014), *Análise da Construção de Túneis com Tuneladoras.* Dissertation (in Portuguese), FCTUC, Coimbra, Portugal, July 2014

Railsystems (2021), *Immersed Tube Tunnel.* Railsystems.com, 2021

Robbins (2021), *A Legacy of Innovation.* Therobbinscompany.com, 2021

Wikipedia (2021), *Tunnel.* Wikipedia, 2021

Wikipedia (2021a), *Tunnel of Eupalinos.* Wikipedia, 2021

Wikipedia (2021b), *Catacombs of Rome.* Wikipedia, 2021

Wikipedia (2021c), *Malpas Tunnel.* Wikipedia, 2021

Wikipedia (2021d), *Thames Tunnel.* Wikipedia, 2021

Wikipedia (2021e), *Holland Tunnel.* Wkipedia, 2021

Wikipedia (2021f), *Tunnel Boring Machine.* Wikipedia, 2021

Wikipedia (2021g), *List of Longest Tunnels.* Wikipedia, 2021

Wills M (2015), *The Channel Tunnel Is a Product of Politics, not Engineering.* Wikimedia Commons, December 2015

Chapter 2

TBM tunnels

Abstract

This chapter presents the main aspects of the tunnel construction using the TBM technique.

The main events in the development of TBMs is presented, as well as the basic aspects related to the construction of tunnels with the TBM technique.

The most significant types of TBMs, according to the type of ground to excavate, are presented: hard rock TBMs (Gripper TBMs, etc.), soft or fractured rock TBMs (Single and Double shield TBMs), soft ground TBMs (EPB and Mixshield TBMs) and heterogeneous ground TBMs (Multi-Mode TBMs), as well as TBMs for urban or undersea tunnels.

Reference is also made to specific issues, such as the thickness of the precast segments, the efficiency index of TBM road tunnels and the construction costs of the TBM tunnels.

2.1 INTRODUCTION

A TBM ("Tunnel Boring Machine") is a "moving factory", a machine that excavates the ground while (in general) placing precast lining segments around the tunnel surface, in order to form the external wall of the tunnel.

TBMs build tunnels, usually with a circular cross-section, boring through all kinds of ground, from hard rock to sand. Excavated diameters can range from 1.0 meter (micro-TBMs) to 17.6 meters to date. Large diameter TBM tunnels are generally considered those over 10 meters excavated diameter.

Among other advantages, TBMs allow to limit the environmental impact to the surface, namely disturbance in the surrounding ground, and produce a smooth tunnel wall. This significantly reduces the cost of the tunnel lining and makes them suitable for use in highly urbanized areas.

The main disadvantage is the upfront cost. TBMs are expensive and can be difficult to transport to the site. However, the longer the tunnel, the lower the relative cost of TBMs compared to alternative methods.

Assuming that TBMs operate successfully, tunnelling with TBMs is very efficient (cost saving) and results in shortened construction times than alternatives. Thus, the TBM technique is nowadays largely the most common in tunnel construction and the most promising for the future.

According to tunnelling tradition, TBMs are always given a name, a female name, without which they don't start working. This tradition is carried out throughout the world as a sign of good luck for the project ahead.

DOI 10.1201/9781003144861-3

2.2 DEVELOPMENT OF THE TBMs

The TBM ("Tunnel Boring Machine") concept is based on a cyclic excavation in which, as a rule and depending on the type of machine, immediate support is provided to the excavated ground.

The cutterhead of the machine is pushed against the excavation front with the aid of hydraulic jacks, while the cutting wheel excavates the mass, making use of the cutting tools installed on it. After the jacks reach their maximum extension, the cutter head stops and the jacks are retracted to their original position.

These jacks act directly on the final support placed immediately behind the machine, still inside the shield of the machine (with the exception of machines that do not carry out the placement of the final support, in which the hydraulic propulsion jacks are connected to claws that act against the walls of the excavated massif). The conclusion of this process thus closes an excavation cycle, also known as excavation advance.

However, to get here a long way had to be taken.

As mentioned in Chapter 1, the TBMs originate from the tunneling shield developed by the French engineer Marc Brunel, a concept he patented two centuries ago, in 1818, and have used for the excavation of the Thames Tunnel under the River Thames in London, UK, which started in 1825, allowing for the first time to build a tunnel under a navigable river.

However, this was only the invention of the shield concept and did not involve the construction of a complete tunnel boring machine, the excavation still had to be carried out by the then conventional tunnelling methods.

As also mentioned, the first TBM (tunnel boring machine), designated as such, was the "Mountain Slicer", developed by the Belgian engineer Henri-Joseph Maus in 1845 for the drilling of the railway Fréjus tunnel, connecting France to Italy, across the Mediterranean Alps.

Built in 1846 in an arms factory near Turin, it consisted of more than 100 percussion drills mounted in the front of a locomotive-sized machine, mechanically power-driven from the entrance of the tunnel. The revolutions of 1848 in Europe affected the funding and the tunnel was completed more than twenty years later, in 1870, by using less innovative methods.

In 1853, in USA, a TBM was also built to be used in the construction of the Hoosac Tunnel in northwest Massachusetts. Made of cast iron, it was known as "Wilson's Patented Stone-Cutting Machine", after inventor Charles Wilson. It drilled 3 meters into the rock before breaking down, being the tunnel eventually completed more than 20 years later, and as with the Fréjus Tunnel, by using less ambitious methods.

However, Wilson's machine anticipated modern TBMs in the sense that it employed cutting discs, like those of a disc harrow, which were attached to the rotating head of the machine. In contrast to traditional drilling and blasting, this innovative method of removing rock relied on simple metal wheels to apply a transient high pressure that fractured the rock.

Also in 1853, the American Ebenezer Talbot patented a TBM that employed Wilson's cutting discs, although they were mounted on rotating arms, which in turn were mounted on a rotating plate.

In the 1870s, the English John D. Brunton built a TBM employing cutting discs that were mounted eccentrically on a rotating plate, so that the cutting discs would travel over almost all of the rock face that was to be removed.

The first TBM that tunnelled significant distances was invented in 1863 and improved in 1875 by British Army officer major Frederick Beaumont (figure 2.1) (Wikipedia 2021).

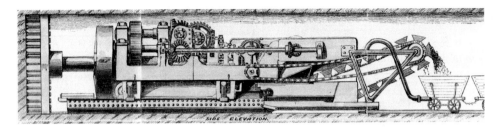

Figure 2.1.   The Beaumont machine

The Beaumont machine was further improved in 1880 by another British Army officer, Major Thomas English (Dennis 2019). The cutting head consisted of a conical drill bit behind which were a pair of opposing arms on which cutting discs were mounted.

In 1875, the French National Assembly approved the construction of a tunnel under the English Channel and the British Parliament allowed a trial to be carried out, after which Major English's TBM was chosen for the project on the English side. From June 1882 to March 1883, the machine excavated through chalk, a total of 1.84 km.

On the French side, a similar machine, developed by the French engineer Alexandre Lavalley (who was also a Suez Canal contractor) also drilled 1.67 km of the tunnel, starting from Sangatte.

However, despite this success, the Cross-Channel tunnel project was abandoned in 1883 after the British military raised fears that the tunnel might be used as an invasion route.

During the late 19th and early 20th century, inventors continued to design, build and test TBMs in response to the growing need for tunnels of all kinds, for railways, roads, subways, sewers, water supplies, etc.

TBMs employing rotating arrays of drills or hammers were in the meantime patented. TBMs that resembled giant saws were also proposed. Other TBMs consisted of a rotating drum with metal tines on its outer surface, or a rotating circular plate covered with teeth, or revolving belts covered with metal teeth.

However, all of these TBMs proved to be expensive, heavy and unable to excavate hard rock, so interest in TBMs has declined. Nevertheless, the TBM development continued in potash and coal mines, for example, where the rock was softer.

In the 1950's, there was an authentic revolution in TBM technology, when the American engineer James S. Robbins was consulted to excavate through a shale geological formation in Colorado, USA, the most difficult to excavate then.

So Robbins developed a machine based on the same principle as the "Mountain Slicer", which was able to excavate about 50 meters in 24 hours.

What made his invention so efficient was to turn the tunnelling head rotating, as anticipated by Charles Wilson. Nowadays it is according to this principle that TBMs work, having evolved in accordance to technological advances and the technical circumstances for which they are requested.

During the 1960s and 1970s, a new revolution in TBMs occurred, with the development in Japan of the so-called "closed-face" machines, which can provide immediate frontal support; first the Slurry machines (improved in the 1980s given rise to Mixshield TBMs) and then the EPB ("Earth Pressure Balance") machines, which have allowed to extend the field of application of TBMs to practically all types of ground.

In the following decades many other innovations have been introduced in TBM technology, with the development of innovative machines, some of which not well succeeded. A curious case is the so-called multi-face TBMs, shielded machines with more than one cutting head, to bore multiple galleries simultaneously (figure 2.2) (Herrenknecht 2019).

Figure 2.2.   Double-face TBM

Among the most recent advances in TBM technology are multi-mode TBMs, which can operate in different modes, allowing to face changeable conditions along the tunnel alignment. The maximum exponent are the so-called Variable Density Multi-Mode TBMs, which can work in four different modes, in open and closed mode and as EPB or Mixshield.

Section 2.4 provides information on the most significant types of TBMs currently in use.

2.3 CONSTRUCTION OF TUNNELS WITH TBMs

TBMs are quite complex equipment, reaching lengths in general over 100 meters (figure 2.3).

They are very powerful machines, achieving very high excavation rates, depending on the type of machine and on the specific ground conditions, becoming common nowadays excavation rates of 100 to 150 meters per week in soil and of more than 200 meters in rock, even in large diameter tunnels. There are recent examples of excavation rates of more than 1200 meters of tunnel per month in rock (Isaman 2019).

The execution of a TBM tunnel begins with the installation of the machine on the ground. Usually, a shaft (or launch box) is created at the beginning of the tunnel, where the different parts of the machine are introduced and assembled.

Figure 2.3.   Schematic view of a TBM

When the TBM reaches its destination, the process is inverted; the machine is progressively disassembled and the different parts are removed, usually, also through a shaft created at the destination point.

As mentioned, the construction of tunnels with TBMs is based on the principle of cyclic excavation. The cutter head in the front of the machine is pushed against the excavation front, using hydraulic jacks, while this wheel rotates and excavates the massif, making use of the excavating tools installed on it. When the jacks reach their maximum extension, the cutterhead stops and the jacks are retracted to their original position.

These jacks act directly on the final support placed immediately behind the machine, with the exception of machines that do not carry out the placement of the final support (Open-face TBMs), in which the hydraulic propulsion jacks are connected to grippers that act against the surface of the excavated tunnel. The completion of this process thus closes an excavation cycle, also known as excavation advance.

TBMs are of different types, according to the conditions of the ground to be bored.

On hard rocks the Open Gripper TBMs are used, in which the cutter head is pushed against the front of the tunnel by hydraulic jacks attached to the excavation wall. In brittle rock formations or soft rock the Single Shield or the Double Shield TBMs are most suitable, using the lining to move the machine forward.

On soft ground, shielded machines with control of pressure on the front are necessary. In the EPB type TBMs the excavated material is removed by a screw conveyor, being moved out of the tunnel by a belt conveyor. In heterogeneous ground, particularly with high water pressure, is more appropriate the use of Slurry (or Mixshield) TBMs, in which a suspension of bentonite is injected on the front of excavation, which mixes with the excavated soil, being the mixture pumped out of the tunnel through a piping network.

Section 2.4 gives more detailed information on the different types of TBMs.

TBMs make use of different cutting tools, depending on the characteristics of the ground to be excavated (figure 2.4). In rocks and soils cutter discs made of high-strength steel are used. As these tools often break, their replacement should be simple and quick.

In heterogeneous ground, in addition to the cutter discs, other types of cutting tools are used for specific purposes, such as cutting knives, scrapers and rippers. They are also made of high-strength steel, sometimes with chromium carbide or tungsten carbide caps, to increase their resistance.

On shielded machines, as excavation progresses, the erector mounts the precast segments (lining) so that the tunnel wall is built immediately behind the shield, which is pushed forward by the hydraulic jacks, which press against the last mounted tunnel ring.

Rock and soil		Soil and heterogeneous ground		
Single cutter discs 	Double cutter discs 	Cutting knives 	Scrappers 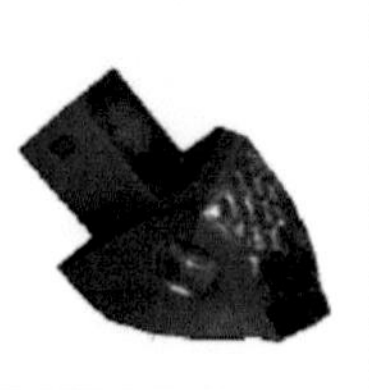	Rippers 

Figure 2.4.   Cutting tools used in TBMs

The rear end of the shield is provided with a sealing in the form of strong brushes through which grease is forced, in order to prevent water and soil from penetrating the space between the shield and the tunnel wall.

The diameter of the excavated tunnel is usually 20 to 40 cm larger than the diameter of the lining, in order to guarantee space for the thickness of the shield and the sealing, and for the inevitable deviations in the conduction of the machine.

At the rear of the machine, inside the finished part of the tunnel, there is the backup system (trailing decks) (see, figure 2.3), that house the support mechanisms, such as the system for muck removal (belt conveyors or slurry pipelines), electrical systems, dust removal, ventilation and mechanisms for transport of precast segments, and usually also the control room of the TBM (figure 2.5).

Figure 2.5.   Control room of a TBM

The final support (lining) of the tunnel is made through rings of precast segments that will form its outer wall (figure 2.6). Longitudinal joints are generally straight, although sometimes other shapes are used (figure 2.7) (Isaman 2019, adapted).

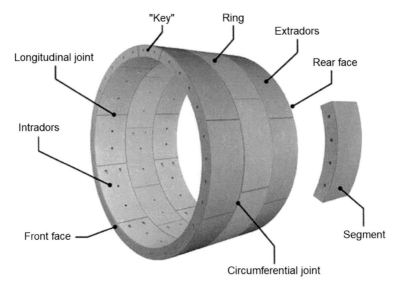

Figure 2.6.   The lining of TBM tunnels

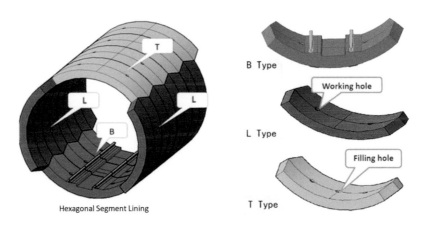

Figure 2.7.   Special lining with hexagonal segments

Precast segments are usually made of high-strength concrete (C45, or higher), reinforced with steel bars or fibers (steel or synthetic), generally, 1.5 to 2.5 meters wide. In the longitudinal and circumferential faces, they are provided with sealings to ensure the tightness of the joints.

The number of precast segments in each ring will be the appropriate to form complete circles with pieces of a given weight, according to the capacity of the handling equipment.

Tunnels of medium size generally have 6 to 8 segments per ring, but in large diameters its number can be of 10 or even more. In each ring there is usually a small segment called "key", wedge shaped, which is the last segment to be mounted. Sometimes it is also adopted a special segment placed at the bottom of the ring, which incorporates the filling at the base of the tunnel.

The precast segments are positioned on the ring through the erector that moves them usually through a suction system. The positioning of the segments on the ring usually follows a mirror lay-out, alternating their position between the right and the left side.

After the completion of a ring, the precast segments are connected together by curved screws and longitudinal keys, to ensure their stability and waterproofing, which are usually removed later. The closing of a ring represents the end of an excavation cycle.

To provide an adequate transfer of efforts between the ground and the lining, the gap between them is filled with injections of cement grouts, using equipment installed in the back-up of the TBM.

In tunnels with shallow cover, a temporary landfill is sometimes installed before the passage of the TBM, to avoid the uplift of the machine.

Nowadays, great attention is being paid to the handling of the excavation products, in order to limit environmental impacts, thus, places with adequate conditions must be found in the tunnel location area. In tunnels excavated in rock, the remains of rock should be processed and reused as much as possible, namely as aggregate in concrete production.

2.4   TYPES OF TBMs

2.4.1   General

Currently, several types of TBMs are available, namely for the construction of traffic tunnels. In choosing the machine for a specific project, it has to be taken into account several parameters such as:

- elementary natural constraints;
- physical parameters such as overall assessment of quality, discontinuities, alterability, water existence and is chemistry;
- mechanical parameters such as the hardness of the soil or rock, deformability and liquefaction potential;
- hydrological parameters;
- design parameters such as size and shape, vertical or horizontal alignment;
- settlements sensibility;
- anomalies in the soil, as natural or artificial obstacles and eventually other parameters such as abrasiveness, soil propensity to stick, the mass/machine friction and the gas presence.

TBMs can be of open-face type, or of shielded type with or without front support. Sophisticated TBMs can combine different characteristics, in order to bore across alignments with different types of ground.

Table 2.1 summarizes the main types of TBMs currently in use. Below are the most significant characteristics of each of these types of machines, as well as some variants, in order to better adapt to the specificities of the ground to be bored.

Table 2.1. Main types of TBMs

Type of ground	TBM type	Type of shield
Hard rock	Open-Face (Gripper) TBM Main-beam TBM Kelly-drive TBM	No shield/Partial shields: roof shield/combination of roof shield and lateral shield/cutterhead shield
	Reamer TBM (TBE)	No shield
Fractured rock	Single Shield Double Shield	Peripheric shield Double peripheric shield
Soil	Earth Pressure Balance (EPB) Slurry Shield (SS)/Mixshield	Peripheral shield and frontal earth pressure Peripheral and front slurry shield
Heterogeneous ground	Multi-mode TBM	EPB/OpenFace; OpenFace/SS; EPB/SS Variable Density TBM (EPB/OpenFace/SS)

2.4.2 Hard rock TBMs

In hard rock (sound rock or with very low alteration degree) and without water, open-face TBMs (completely open or partially shielded) are usually used.

The open-face TBMs are named **Open TBMs** or **Gripper TBM**s. There are also the so-called **Main-beam TBMs** and the **Kelly-drive (Double Gripper) TBMs**, which differ on the type of grippers used.

Hard rock TBMs excavate the rock with disc cutters mounted in the cutterhead. The disc cutters create compressive stress fractures in the rock, causing it to chip away from the tunnel face. The excavated rock (muck) is then transferred through openings in the cutterhead to a belt conveyor, where it runs through the machine to a system of conveyors or muck cars for removal from the tunnel.

In principle, the open-face TBMs have no shield, leaving the area behind the cutterhead open for rock support. To advance, the machine uses a gripper system that pushes against the excavated tunnel walls. Some machines can't be continuously steered while gripper shoes push on the walls, thus they will push forward off the grippers gaining thrust.

At the end of a stroke, the rear legs of the machine are lowered, the grippers and propel jacks are retracted. The retraction of the propel jacks repositions the gripper assembly for the next boring cycle. The grippers are extended, the rear legs lifted, and boring resumes.

So far, this type of machines comprises diameters up to 12.5 meters. Figure 2.8 shows the main components of the open-face TBMs (Herrenknecht 2020).

Although, in principle, open-face TBMs do not have a shield, small shields are sometimes installed to protect the machine's organs and the operators from occasional falling debris.

Different types of shields are used, according to the type of protection conferred: the roof shield, a combination of a lateral shield and a roof shield, and the cutterhead shield.

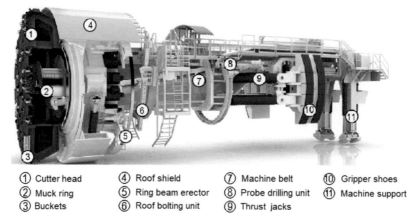

① Cutter head	④ Roof shield	⑦ Machine belt	⑩ Gripper shoes
② Muck ring	⑤ Ring beam erector	⑧ Probe drilling unit	⑪ Machine support
③ Buckets	⑥ Roof bolting unit	⑨ Thrust jacks	

Figure 2.8. Scheme of an open-face TBM (with roof shield)

The open-face TBMs does not install concrete segments behind as other machines. Instead, the rock is held up using ground support methods such as ring beams, rock bolts, steel straps, ring steel and wire mesh, protected with shotcrete.

Although not much used, for this type of ground there is also the **Reamer TBMs**, also called TBEs ("Tunnel Boring Extender"), which are similar to the Open TBMs presented above, except that the cutterhead is pulled against the excavation face instead of being pushed.

This movement is performed through a traction unit with grippers installed in the front of the machine, that act inside a pilot hole open in the excavation front, which is previously excavated by a common TBM. Basically, a Reamer TBM widens a pre-existing hole.

2.4.3   Soft or fractured rock TBMs

In soft or fractured rock (without water), shielded hard rock TBMs are used, which erect precast concrete segments to support unstable tunnel walls behind the machine. The other componentes of the machines are similar to those of the Open TBM.

Two types of machines are available: the **Single Shield TBMs** and the **Double Shield TBMs**. So far, they comprise diameters up to 14.0 meters and 12.5 meters, respectively.

Single Shield TBMs
Single Shield TBMs are used in fractured ground, as they push against the concrete lining segments (figure 2.9) (Herrenknecht 2020).

Hydraulic jacks radially arranged aplly the necessary thrust forces. Those installed at the bottom of the machine need to apply greater forces due to the friction generated between the shield and the massif.

An important detail is that in these machines the diameter of the cutterhead is slightly larger than that of the shield and its axis is positioned slightly above the axis of the machine, in order to create an "overcut" to prevent it from jamming against the massif, in the case of squeezed (plastic) rock.

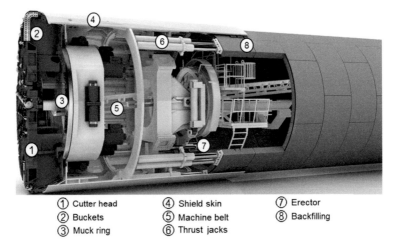

① Cutter head   ④ Shield skin   ⑦ Erector
② Buckets   ⑤ Machine belt   ⑧ Backfilling
③ Muck ring   ⑥ Thrust jacks

Figure 2.9.   Scheme of a single shield TBM

Double Shield TBMs

Double Shield TBMs are considered to be the most sophisticated machines for boring in rocky ground, as they combine the operating principles of the Open TBMs and those of the Single Shield TBMs, allowing high excavation rates. The cutterhead is identical to that of the Single Shield TBMs (figure 2.10) (Herrenknecht 2020).

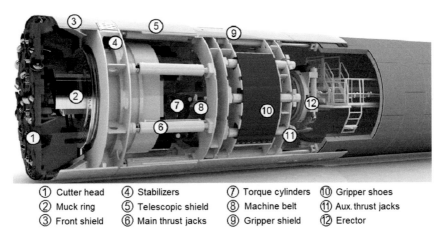

① Cutter head   ④ Stabilizers   ⑦ Torque cylinders   ⑩ Gripper shoes
② Muck ring   ⑤ Telescopic shield   ⑧ Machine belt   ⑪ Aux. thrust jacks
③ Front shield   ⑥ Main thrust jacks   ⑨ Gripper shield   ⑫ Erector

Figure 2.10.   Scheme of a double shield TBM

The Double Shield TBMs function in two modes; in stable ground the rear shield (Gripper shield) grip to the tunnel surface to advance. In unstable, fractured ground, the thrust is shifted to radially arranged thrust jacks that push the front shield (telescopic shield) against the tunnel segments rings behind the shield (through the rear shield). This keeps the thrust forces from impacting the tunnel's fragile walls. Telescopic and Gripper shields are articulated.

In competent massifs, these two systems can work alternately, thus there are no excavation cycles (excavation and erection of segments are simultaneous) and the machine can work continuously without stops, which allows it to achieve high rates of excavation.

2.4.4   Soft ground TBMs

The excavation of tunnels in soft ground (earthy massifs or highly fractured rock) is one of the most complex tasks in underground works.

Unlike excavation in hard rock, the main difficulty here is not related with the excavability of the massif, but to the stability of the excavation front, especially when there is water around (figure 2.11) (Herrenknecht 2020).

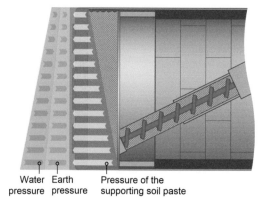

Water   Earth   Pressure of the
pressure   pressure   supporting soil paste

Figure 2.11.   Actions in front of a TBM on soft ground with water

So, in addition to the tendency of the soil to stick to the cutting head and the possible existence of gases when the massif is rich in organic materials, there is a need to provide frontal and lateral support, before the final lining is placed, especially in urban areas, where the coverage is usually low and there is the risk of settlements.

For these ground conditions, two main types of TBMs exist: **EPB** ("Earth Pressure Balance") TBMs and **SS** ("Slurry Shield") TBMs. The propulsion of these types of machines is assured, as in the Single Shield TBMs, by thrust jacks radially disposed, that press against the last installed ring of precast segments.

Earth Pressure Balance (EPB) TBMs
For soft and cohesive soils the EPB TBMs are suitable. They also work well in rock ground with water. They get the name because they use the excavated material to balance the pressure at the tunnel front.

The spoil excavated by the cutting wheel (cutter head) is stored inside the excavated chamber (plenum), from which it is extracted through the screw conveyor (Archimedes screw). The necessary pressure is maintained in the cutting wheel by controlling the rate of extraction of spoil through the screw conveyor and the advance rate (figure 2.12) (Herrenknecht 2020).

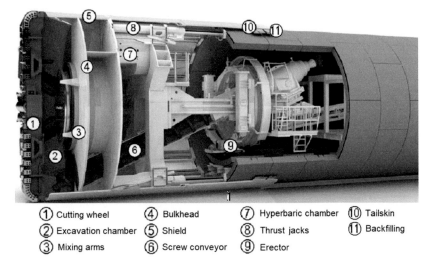

① Cutting wheel	④ Bulkhead	⑦ Hyperbaric chamber	⑩ Tailskin
② Excavation chamber	⑤ Shield	⑧ Thrust jacks	⑪ Backfilling
③ Mixing arms	⑥ Screw conveyor	⑨ Erector	

Figure 2.12.   Scheme of an EPB TBM

Water and additives like polymers and foam are usually injected under pressure in front of the face to increase the stability of the ground ahead and the workability of the excavated material that is the spoil remains sufficiently cohesive to form a plug in the screw conveyor to maintain pressure in the cutting wheel. The balance in the excavation chamber is continuously monitored using earth pressure sensors.

The homogeneity of the mixture is assured by the rotation of the mixing arms inside the excavation chamber. These machines may also be equipped with special tools to crush higher blocks. From the screw conveyor, the paste (muck) is deposited in a conveyor belt to be transported in solid state out of the tunnel, using special vehicles or a conveyor belt system.

These TBMs need the placing of a final support (lining), which is mounted inside the shield, which is provided with a membrane that ensures the sealing between the inside and outside of the machine. This sealing is achieved by means of wire brushes that are attached to the shield (tail skin) and act against the precast segments.

The cutting tools used depend on the type of massif to be excavated. Thus, the cutting wheel, in addition to disc cutters, uses cutting tools that are more suitable for excavating in soil, such as cutting knives and scrappers (see, figure 2.4).

These machines are equipped with a hyperbaric chamber, through which operators pass to adapt their bodies when they need to enter the excavated chamber (which is under pressure) to change the cutting tools, which happens quite often.

EPB machines are used on ground up to 7 bar of pressure. So far, they comprise diameters up to 17.5 meters.

Slurry Shield (SS) TBMs

On soft, heterogeneous ground with very high water pressure or where ground conditions are granular (sand and gravel) so that a plug could not be formed in an Archimedes screw, Slurry Shield TBMs are required.

In these machines, also known as **SFM (Slurry Face Machines)** and **Mixshield TBMs**, the cutting head is filled with pressurised slurry, usually bentonite clay, which applies hydrostatic pressure to the excavation face to provide the necessary support to stabilize the excavation front and to prevent uncontrolled tunnelling penetration.

The slurry also acts as a transport medium by mixing with the excavated material before being pumped out of the cutting head back to a slurry separation plant, usually outside of the tunnel.

As in the EPB machines there is a sealing mechanism between the shield and the lining, to prevent water from entering the machine, which is more sophisticated, since, in general, it withstand higher pressures.

The cutting head is also similar to that of an EPB, as well as the excavation tools. However, as in non-cohesive and heterogeneous soils, large stones or blocks can be expected, these machines are often equipped with a jaw crusher to crush them to a conveyable size so that hydraulic removal can be carried out.

The operation of these machines is, in general, more complex than the EPB machines, due to the more complex pressurization system, which obliges to the construction of a complex piping network, as well as a slurry preparation and separation plant.

In **Mixshield** machines, the excavation chamber is divided by the so-called submerged wall, provided with a communication opening at the bottom, the front section being completely filled by the mixture in order to promote the support of the excavation front, while at the rear the mixture will up to half the height, the rest being filled space by an air cushion. This allows a better control of the pressure inside the chamber and consequently in the excavation front, allowing to work in high pressures, more than 15 bar (figure 2.13) (Herrenknecht 2020).

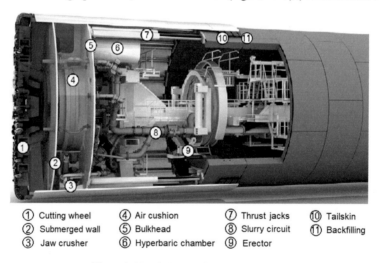

① Cutting wheel	④ Air cushion	⑦ Thrust jacks	⑩ Tailskin
② Submerged wall	⑤ Bulkhead	⑧ Slurry circuit	⑪ Backfilling
③ Jaw crusher	⑥ Hyperbaric chamber	⑨ Erector	

Figure 2.13.   Scheme of a Mixshield TBM

As the EPBs, these machines are equipped with a hyperbaric chamber.

A 17.6 meter diameter Mixshield TBM has already been built, the largest in the world. A 19.25m diameter Mixshield was also designed for the Orlovski Tunnel, a project to be carried out in Saint Petersburg, Russia, but it was not built.

Slurry separation plants are multi-stage filtration systems, which remove particles of spoil from the slurry so that it may be reused in the construction process (figure 2.14).

Figure 2.14.   Slurry separation plant

The limit to which slurry can be "cleaned" depends on the particle size of the excavated material. For this reason, slurry TBMs are not suitable for silts and clays as the particle sizes of the spoil are less than that of the bentonite clay from which the slurry is made.

In this case, the slurry is separated into water, which can be recycled and a clay "cake", which may be polluted, is pressed from the water.

EPB versus SS TBMs
As mentioned, the EPB and SS TBMs have their specific field of application, depending on the size of the particles of the ground: clay, silt, sand or gravel and their percentage. Figure 2.15 (Biggart 2011, adapted) shows the PSD (Particle Size Distribution) diagram with the fields of application of each type of machine.

However, as it can be seen, there is a mid-range of ground conditions which is suitable for both types of machines. In these situations, additional information will be needed, namely about the ground permeability and the hydrostatic head, and the distribution of fines along the whole length of the tunnel.

2.4.5 Heterogeneous ground TBMs

Classic tunnelling shielded machines can reach their technical or economic limits with their specific method when they need to drive through highly variable geologies. In fact, along a tunnel alignment, heterogeneous grounds can change from stable rock to soft ground with water and vice-versa, especially in long tunnels.

For these situations, manufacturers have developed multi-mode TBMs, in order to ensure optimum safety and flexibility along the entire length of the tunnel. These machines are flexibly designed in terms of support and excavation methods.

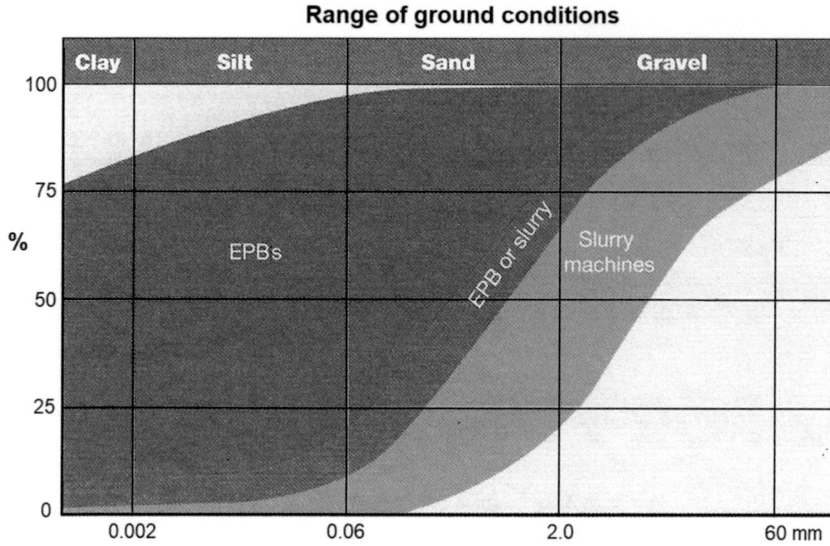

Range of ground conditions

Figure 2.15. The PSD curve for grain size distribution

The tunnelling mode can be adapted to changing ground, requiring relatively short conversion times and incurring low costs. This means that even tunnels with extremely varying geological and hydrogeological conditions can be constructed safely and cost-effectively.

As mentioned, in TBM tunnelling there are basically three different shield types: Open-face Shields, Earth Pressure Balance Shields and Slurry Shields. Each of these methods has its particular range of application.

Technological advances in recent years have enhanced the geological range of the machines, combining different methods, in order to increase the efficiency of the projects. Two variants are possible:

a) Multi-mode TBMs with a modular basic structure, in which the tunnelling method is changed during tunnelling by extensive modifications to individual components.
b) Multi-mode TBMs with components that allow different tunnelling methods, which are all integrated into the machine. Naturally, these machines have a more complex technical design. Different combinations are possible, the following being the most representative.

Combination of EPB and Open-face TBM

EPB Shields with screw conveyor in the invert section can be driven in conventional closed EPB mode and also in open-mode. The excavation chamber and the screw are in this case only partly filled. In this mode, no active support pressure against the tunnel face is applied and the screw conveyor is used only for removing the excavated material.

Alternatively, a belt conveyor with a retractable muck ring can be installed in the center. This requires additional conversion measures at the cutterhead, and the screw conveyor is partly retracted in the open-mode. The combination of EPB Shield and Open-face TBM can be designed with a modular basic structure and as a version with parallel screw and central belt conveyor.

Combination of Open-face and Slurry TBM

Multi-mode TBMs with open and slurry-supported mode have been used successfully in many projects. The greatest challenge when changing the tunnelling mode is to deal with the different ways of removing the excavated material.

While the slurry-supported shield has a slurry circuit installed, in the open mode the material is removed using a central belt conveyor with retractable muck ring. This means that both conveying systems must be installed on the TBM and the back-up system in such a combination. Modular and integrated machine concept solutions are available.

Combination of EPB and Slurry TBM

Both modes of this combination work with a closed pressure system to actively support the tunnel front. The main differences consist in the composition and characteristics of the support medium, the conveying of the excavated material and the excavation chamber design.

However, parallel installation of screw conveyor and slurry circuit in the invert section of the excavation chamber is possible in large diameter machines only (more than 9 meters). For smaller diameters, the tunnelling mode requires time and incurs costs. This is particularly true if an additional stone crusher is needed in front of the slurry circuit intake screen.

The special case of the Variable Density TBM

For tunnel alignments with highly changeable ground conditions, a special type of Multi-mode TBM, the **Variable Density Multi-Mode** (VDMM) TBM was developed, which applies a unique tunnelling technology (figure 2.16) (Herrenknecht 2020).

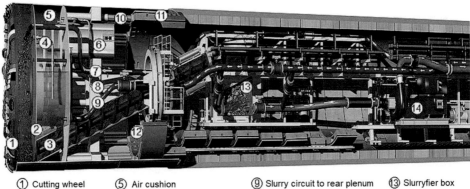

① Cutting wheel	⑤ Air cushion	⑨ Slurry circuit to rear plenum	⑬ Slurryfier box
② Submerged wall	⑥ Hyperbaric chamber	⑩ Thrust jacks	⑭ Pump slurry extraction
③ Screw conveyor	⑦ Slurry circuit to front plenum	⑪ Tailskin	
④ Connection pipes	⑧ Compressed air circuit	⑫ Erector	

Figure 2.16. Scheme of a multi-mode TBM with variable density technology

Without major mechanical modifications, the machine can switch between four different tunnelling modes, directly in the tunnel. The basic concept makes it possible to change between slurry support (Mixshield), earth pressure support (EPB) and open-mode within the same tunnel alignment, by undertaking a range of conversions.

This means that geological and hydrogeological changes along the tunnel alignment can be managed with great flexibility, allowing for high excavation rates in safety.

2.4.6   TBMs for urban and near surface tunnelling

The construction of TBM tunnels in urban environment usually requires boring near surface (low cover), as it allows less deep access shafts or shorter access ramps. However, it should be guaranteed that the ground surface is not undisturbed, which means that ground subsidence has to be avoided. The usual way to do this on soft ground is to maintain ground pressure during and after the tunnel construction.

However, it should be borne in mind that sometimes there are difficulties in doing so, particularly in varied strata, for example, boring through an area where the upper part of the tunnel face is wet sand and the lower part is hard rock.

EPB and **SS** type TBMs are those usually used in these situations. Both types of machines are capable of reducing the risk of surface subsidence and voids, if operated properly and ground conditions are well documented. The use of sensors to control the movements of the soil and the main utilities (buildings, etc.) is imperative.

As mentioned above, in tunnels with shallow cover (in particular, underwater), it may be necessary to install a temporary landfill, to avoid the uplift of the machine.

Before tunnelling, it will be necessary to carry out a detailed survey of possible existences in the ground, such as other tunnels, utility lines and deep foundations, often not well documented in the existing information. In addition, the project must include measures to mitigate any detrimental effects on other infrastructure.

2.4.7   TBMs for undersea tunnels

The building of undersea TBM tunnel projects present significant challenges, both geotechnical and operational.

With regard to geotechnical challenges, although they do not differ in magnitude from large projects inshore, undersea tunnelling poses a unique combination of challenges, due to the geologic variability. In addition, in subsea tunnels the geological investigation and mapping is usually more difficult to carry than in land.

Tunnels crossing ocean straits, lakes and rivers inevitably cross various geological formations, from superficial deposits, which require TBMs with active support, to bedrock formations which require TBMs for hard-rock excavation, with potentially high water pressures (specially in deep alignments).

Another important issue are the man made obstacles underground, especially in harbour or strait areas, such as unchartered piles, drill casings or objects sunken into the sea that can severely damage the excavating machines.

With regard to operational challenges, important issue is maintenance of the TBM, since maintenance work on the cutterhead, under high pressures, may be necessary. In long tunnels this means greater logistic requirements, as well as regarding mucking, ventilation and transports. Regarding the high pressures, pressures above 7 bars, rule out the EPB machines. Additionally, as those environments are usually unlimited reservoirs of water, tunnels can be flooded in short time, so they represent a much higher risk than in land.

The use of adequate technologies is crucial for the success of the project. Multi-mode TBMs on their appropriate combinations (see, Section 2.4.5) allow excavation on open-mode over long sections, resulting in high excavation rates. When there is too much water inflow, the excavation mode can be quickly changed to a closed mode, until another stable formation is reached.

Accessible cutterheads, with their large hollow arms, can be accessed from the TBM under atmospheric pressure. Saturation diving thus allows divers to make repairs at pressures up to 45 bar (Duhme 2017). In addition, although the replacement of the cutting tools still requires human intervention, the inspection works can nowadays be fully automatized.

Seismic exploration and other geological survey techniques allow the reliable detection in advance of obstacles, such as boulders, piles, fault zones or similar geological phenomena, which allow to simplify the geological surveys that are carried out ahead.

2.5 THE RACE FOR THE MEGA TBMs

Mostly, from the beginning of the 2000s, there was a kind of race between the main TBM manufacturers to be the suppliers of machines with the largest diameters.

In 2003, the Westerschelde Tunnel was opened to traffic in the South of Netherlands (see, Section 4.3.2), a dual-tube roadway tunnel, built by two Mixshield type TBMs, "Sara" and "Suzanna" supplied by Herrenknecht AG, each measuring 11.33 meters in diameter, the world record at the time.

Between 2002 and 2005, an even larger TBM, a slurry-type TBM supplied by NFM Technologies, bored the mono-tube railway Groeneheart Tunnel in the HSL-Zuid, in The Netherlands (see, Section 4.2.4), with a diameter of 14.87 meters, making the world record for TBMs at the time and still today the world record for railway tunnels.

This TBM was surpassed by the EPB TBM "Dulcinea", supplied by Herrenknecht AG, which, between 2005 and 2008, built the North tube of the dual-tube South By-Pass Tunnel on the M30 ring road in Madrid, Spain (see, Section 4.3.2), with a diameter of 15.20 meters.

In 2006, the world record went to the two Mixshield TBMs that, between 2006 and 2009, built the Changjiang dual-tube tunnel in Shanghai, China (see, Section 4.3.2), with diameters of 15.43 meters, the main components being supplied by Herrenknecht AG and assembled by STEC (Shanghai Tunnel Engineering Co), its partner in Shanghai, China.

Between 2011 and 2013, "Martina", an EPB TBM, supplied by Herrenknecht AG, 15.60 meters diameter built the two tubes of the roadway Sparvo Tunnel, on the A1 highway, in the Appenins, near Firenze, Italy (see, Section 1.2), becoming the new world record.

In 2013, "Bertha", a 17.50 meters diameter EPB TBM supplied by Hitachi Zosen Corporation, became the world's largest TBM in the world, starting to excavate the mono-tube SR99 roadway tunnel, in Seattle, USA, which job ended in 2017 (see, Section 4.3.3).

In 2015 the world record rose to 17.60 meters from the Mixshield TBM supplied by Herrenknecht AG and assembled at its factory in Nansha (near Hong Kong), China, which in just a few months built the 0.6 km long Northern Connection of the Tuen Mun-Chek Lap Kok Tunnel in the TM-CLK Link in Hong Kong (see, cover and figure 2.17) (Herrenknecht 2020). Subsequently, the shield of the machine was changed to a diameter of 14.0 meters to bore the second tube of the Southern section of the referred dual-tube tunnel, which, like its "sister" (also 14.0m in diameter) ended the work in 2019.

Figure 2.17.   The 17.60m diameter TBM shield for the Tuen Mun-Chek Lap Kok Tunnel in the TM-CLK Link in Hong Kong, China, currently the largest in the world

In 2017, a 15.87 meters diameter EPB TBM, also supplied by Herrenknecht AG, started the construction of the mono-tube Santa Lucia road tunnel, 7.5 km long, also on the A1 highway, in the Appenins, Italy (similar to Sparvo Tunnel), becoming the largest TBM operating in Europe (and the third in the world), being the work competed in 2020.

It should also be noted that in 2011 Herrenknecht AG completed the design of a 19.25 meters diameter Mixshield TBM for the mono-tube roadway Orlovski Tunnel in Saint Petersburg, Russia (see, Section 4.3.3), but it was not built.

2.6   THICKNESS OF THE PRECAST SEGMENTS OF TBM TUNNELS

The thickness of the precast segments of the outer wall (lining) of the TBM tunnels depends on several factors, mainly the strength characteristics of the ground, the surrounding acting stresses (namely the water pressure), the seismicity of the site and the thrust forces applied by the TBM.

Figure 2.18 shows the internal diameter (Di) and the thickness of the precast segments (t) of several TBM tunnels built around the world in a wide range of conditions.

It can be seen that, in general terms, these parameters fit relatively well to the relationship: $t = Di/25$.

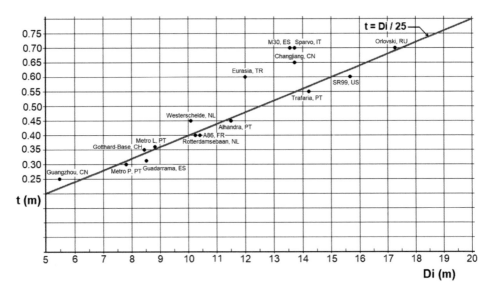

Figure 2.18. Relationship between the thickness of the precast segments and the inner diameter of TBM tunnels

However, under special conditions, namely high water pressure (above 6 bar), large overburdens or strong seismic conditions, precast segments with greater thicknesses may be necessary.

2.7   EFFICIENCY INDEX OF THE CROSS-SECTION OF TBM ROAD TUNNELS

Despite the aforementioned advantages, TBM tunnels, as their cross section is circular, there is no full use of the area of its internal cross section, particularly in the case of road tunnels.

Thus, for each TBM road tunnel the respective **efficiency index**, defined as the ratio between the roadway cross-sectional area: clear height x total width dedicated to roadway purposes (roadway lanes + edges and/or emergency lane, when existing) and the internal cross-sectional area of the tunnel, should be established, which will be a useful indicator to assess the cost/benefit ratio of the solution.

The efficiency index of the TBM road tunnel should be as high as possible, although it generally assumes values in the range 0.40 to 0.50. In the following chapters, significant TBM road tunnels are presented, with indication of the respective efficiency index.

2.8   CONSTRUCTION COSTS OF TBM RAILWAY AND ROAD TUNNELS

Construction costs for TBM rail and road tunnels can vary significantly, depending on several factors linked to the location: geological conditions (which condition the type of machine, excavation rates, etc.), geographic location (urban or non-urban), accessibility (access roads, etc.) and environment protection (need for mitigation measures, etc.), or to intrinsic characteristics of the tunnel, namely, length and diameter, number of tubes (monotube/dual-tube) and nature of traffic (light vehicles, HGV, etc.).

Although it is generally not possible to quantify the influence of each of these parameters, it is always possible to make estimations, based on the costs of tunnels already built, taking into account their specific conditions, or at least indicate average costs.

A useful parameter for making TBM tunnel construction cost estimates is the **tunnel construction unit cost**, defined as the ratio of the overall tunnel construction cost to the excavation volume.

On Table 2.2 and Table 2.3 the construction unit costs of several railway and road tunnels, recently built worldwide in a wide range of local conditions (those for whom it was possible to access comparable data), are presented.

Table 2.2. Construction unit costs of TBM railway tunnels

	Length (Km)	Diameter (m)	Volume (Mm^3)	Cost (M€)	Unit cost (€/m^3)
Perthus (France/Spain)	2 × 8.3	9.90	1.28	300	234
Guadarrama (Spain)	2 × 28.4	9.51	4.03	1220	303
Aconcagua (Chile/Argentina)	2 × 52.5	10.0	8.24	2700	328
Groenehart (Netherlands)	1 × 8.0	14.9	1.39	430	309
Atocha-Chamartin (Spain)	1 × 7.3	11.5	0.76	200	263
Average					**287**

Table 2.3. Construction unit costs of TBM road tunnels

	Length (Km)	Diameter (m)	Volume (Mm^3)	Cost (M€)	Cost (€/m^3)
Westerschelde (Netherlands)	2 × 6.6	11.33	1.33	726	545
Waterview (New Zealand)	2 × 4.0	13.0	0.53	230	434
A86, Paris (France)	1 × 10.0	11.6	1.05	300	286
Sparvo (A1), Florenze (Italy)	2 × 2.6	15.6	0.30	150	302
Changjiang, Shanghai (China)	2 × 7.5	15.43	2.80	1200	430
Rotterdamsebaan, Hague (Netherlands)	2 × 3.8	11.17	0.74	272	367
Nanjing-Weisan (China)	3.5 + 4.1	14.9	1.32	630	477
Average					**406**

Thus for railway tunnels, in 5 cases, construction unit costs between EUR 234 per cubic meter and EUR 328 per cubic meter were obtained, with an average value of EUR 287 per cubic meter of excavation.

As regard to road tunnels, in 7 cases, construction unit costs between EUR 286 per cubic meter and EUR 545 per cubic meter were obtained, with an average value of EUR 406 per cubic meter of excavation.

It can also be seen that, in general terms, the construction unit costs of TBM railway tunnels are about 70% of the construction unit costs of TBM road tunnels.

It is also worth mentioning the tendency of reduction in the construction unit cost with the increase in the length of the tunnels, as well as the monotube tunnels in relation to the dual-tube tunnels.

2.9 CONCLUSIONS

TBM tunnels are tunnels built by TBMs ("Tunnel Boring Machines"), that is, machines that excavate the ground while (in principle) placing precast segments around the tunnel surface (lining), in order to form the outer wall of the tunnel.

TBMs originate from the tunnelling shield created by Marc Brunel in 1818, associated with the rotary cutting head developed by James S. Robbins in 1950. Subsequent developments gave rise to different types of machines, thus today there is available a panoply of TBMs, according to the geological conditions of the ground to be bored: Gripper (Open) TBMs for hard rock; Single Shield and Double Shield TBMs for soft or fractured rock; and EPB ("Earth Pressure Balance") and Slurry Shield (Mixshield) TBMs for soft soil.

In soft soil, TBMs, in addition to excavating the ground and installing the precast segments, provide frontal and lateral support, in order to allow excavation under high pressures, namely from the surrounding water. The choice between these two types of TBMs will depend mainly on the specific characteristics of the soil: EPB TBMs for soft and cohesive soils (or rocky ground with water) and Mixshield TBMs when the soil conditions are granular (sand and gravel) with very high water pressure.

In heterogeneous ground, multi-mode machines are used, which combine the characteristics of the machines mentioned above. The most advanced are the so-called Variable Density Multi-Mode, which can switch between four different tunnelling modes.

The diameter of TBMs have been steadily increasing during the last two decades, with 17.60 meters the current world record.

The thickness of the precast segments of the tunnel wall depends on several factors, namely the conditions of the ground and the thrust forces applied by the TBM. For common situations, it is about 1/25 the inner diameter of the tunnel.

In road TBM tunnels the efficiency index (ratio between the roadway cross-sectional area and the internal cross-sectional area of the tunnel) is a useful parameter to assess the cost/benefit ratio of the solution.

As for TBM tunnel construction costs, a very useful parameter is the tunnel construction unit cost, the ratio between the overall tunnel construction cost and the excavation volume, which can be a good help in making estimates.

REFERENCES

Baraniuk C (2017), *The Channel Tunnel that Was never Built.* BBC Future, August 23, 2017
Biggart A (2011), *Closed Face Decisions – Slurry or EPBM?* Tunnelling Journal, February/March 2011

Dennis G (2019), *Channel Tunnel: The 20th Century's Most Expensive Construction Project – But Worth Every Penny*. Rail Magazine, Issue 878, May 8, 2019

Duhme R (2017), *Designing TBMs for Subsea Tunnels*. The 2017 World Congress on Advances in Structural Engineering and Mechanics (ASEM17), Ilsan, Korea, August 2017

FHWA (2009), *Technical Manual for Design and Construction of Road Tunnels – Civil Elements*. Federal Highway Administration, U.S. Department of Transportation Publication Nº FHWA-NHI-10-034, December 2009

Halim I (2020), *TBM Tunnelling in Mixed Ground Conditions*. Tunnel Business Magazine, June 2020

Herrenknecht (2011), *Pioneering Tunnelling Technology for Underground Transport Systems*. Herrenknecht AG, 2011

Herrenknecht (2021), *The Perfect Technology for your Challenge*. Herrenknecht.com, 2021

Herrenknecht, M (2019), *Innovations in Mechanized Tunnelling since 1970*. ITA-AITES Muir Wood Lecture 2019, April 2019

Isaman M (2019), *Breaking through Tough Ground in the Himalayas: Nepal's First TBM*. Society for Mining, Metallurgy & Exploration, 2019

ITA-AITES (2004), *Recommendations and Guidelines for Tunnel Boring Machines (TBMs)*. ITA-AITES Working Group WG 14 – Mechanized Tunnelling, 2004

Maidl B, Herrenknecht M, Maidl U, Wehrmeyer G (2012), *Mechanised Shield Tunnelling*. Ernst & Sohn, 2nd Edition, Berlin, Germany, 2012

Nôro A M (2014), *Análise da Construção de Túneis com Tuneladoras (Analysis of Tunnel Construction with Tunneling Boring Machines)*. Dissertation (in Portuguese), FCTUC, Coimbra, Portugal, July 2014

PIARC (2011), *Road Tunnels Manual*. World Road Association (PIARC), August 2011

Robbins (2021), *A Legacy of Innovation*. Therobbinscompany.com, 2021

TunnelTalk (2019), *Tracking the World's Mega-TBMs*. TunnelTalk, Mars 2019

Wallis S (2015), *Boundaries Busted for Highway under the Bosphorus*. TunnelTalk, September 24, 2015

Wikipedia (2021), *TBM*. Wikipedia, 2021

Chapter 3

Safety of traffic tunnels

Abstract

This chapter presents the main safety requirements for traffic tunnels (railway tunnels and road tunnels): fire safety, emergency services, evacuation and rescue, etc.

An overview of the most significant existing normative documents, mostly at EU (European Union) level, is presented: Directive 2004/54/CE for road tunnels and the EU Decision 163CE/2008 for railway tunnels.

The most significant technical issues related with the safety of traffic tunnels are analysed: cross-section of railway and road tunnels, emergency exits, ventilation systems, fire fighting systems (in particular, Water Based Fire Fighting Systems) and safety in operation concepts.

3.1 INTRODUCTION

Safety of traffic tunnels has received significant attention in recent time as a result of major accidents that led to fatalities and considerable economic damage. In fact, due to the enclosed nature of a tunnel, even the occurrence of a relatively minor incident has the potential to develop into a catastrophe as well as cause significant damage to the tunnel structure.

Thus, international institutions and countries that have tunnels have developed recommendations and guidelines or adopted regulations for the design, construction, operation, maintenance and intervention of the emergency services in tunnels.

Regarding safety conditions of railway tunnels, the matter is treated by the UIC (International Union of Railways) on the UIC Code 779-9: "Safety in Railway Tunnels" (UIC 2003). In turn, EU countries are bounded by the EU Decision 163CE/2008: "Safety in Railway Tunnels" (EU 2008), which is mandatory to tunnels more than 1.0 km in length.

As regard the safety conditions of road tunnels, EU countries are subjected to Directive 2004/54/CE: "Safety in Roadway Tunnels" (EU 2004) that prescribes arrangements to be implemented in order to ensure safety of users in tunnels longer than 0.5 km that are part of the Trans-European Road Network.

The main aspects of these documents are presented below, as well the technical solutions usually adopted to satisfy their requirements. It should be noted that these documents are intended to prescribe minimum requirements, so that, in specific situations, more demanding measures may be adopted, if justified. Some European countries are also adopting specific criteria on this matter, which in some situations are more demanding than the EU documents.

The American NFPA 130: "Standard for Fixed Guideway Transit and Passenger Rail Systems" (NFPA 2020), also presents guidance and criteria for safety in railway systems, applicable to railway tunnels.

DOI 10.1201/9781003144861-4

3.2 SAFETY OF RAILWAY TUNNELS

Despite the risks, there is a widespread perception that railway tunnels are quite safe, safer than road tunnels, for example. However, risks exist, so international institutions have issued documents in order to harmonize practices between countries.

The UIC Code 779-9 (UIC 2003) is focusing mainly on risk analysis, presenting an inventory of the risks associated with the operation of the railway tunnels, with a particular emphasis on derailment, train crash and, above all, fire.

The EU Decision 163CE/2008 (EU 2008) supplements the Directive 2001/16/CE on the interoperability of the conventional Trans-European railway system and approves the ETI (Technical Specification for Interoperability): "Safety in Railway Tunnels", which applies to tunnels in both the common and the high-speed railway lines.

On the ETI, safety focuses primarily on providing measures for the event of an accident or fire inside the tunnel. Among other requirements, the ETI prescribes the installation of emergency walkways along each track, at least, 0.70m wide. In the case of dual-tube tunnels it prescribes also the installation of cross-passages with a maximum spacing of 0.5 km and of access galleries to outside, at intervals not exceeding 1.0 km. All of these galleries must be pressurized, to avoid smoke entering. However, there are no ventilation requirements for the traffic galleries, namely in fire situations.

A key issue regarding the safety of railway tunnels is the number of tubes to make up the tunnel: mono-tube tunnels (with double track), or dual-tube tunnels (each with single track).

However, both the UIC Code and the EU ETI do not comment on this matter, so in each case the decision will have to be made based on a risk analysis, in order to comply with safety requirements.

3.3 SAFETY OF ROAD TUNNELS

In the EU Directive 2004/54/CE (EU, 2004) safety of road tunnels is focused on the risk analysis of each case, considering as important issues the geometry and length of the tunnel, and the traffic volume (AADT per lane) and its type (heavy goods, etc.).

Regarding the number of tubes (single or twin tube tunnels) the Directive establishes that it should be based on the projected traffic volume (below or above 10 000 daily vehicles per lane), taking into account aspects such as the percentage of heavy good vehicles, gradient and length.

Some infrastructure measures are prescribed, such as: longitudinal gradient not higher than 5%; emergency walkways (elevated or not) to be adopted aside of each way, if there is no emergency lanes; emergency exits (to outside) spaced 0.5 km maximum; emergency stop areas at distances not more than 1.0 km, if there are no emergency lanes.

Other measures are also foreseen, namely: ventilation systems are compulsory in tunnels over 1.0 km long; a water supply shall be installed in all tunnels; hydrants shall be placed at intervals not exceeding 250 meters and a Control Centre shall be installed in tunnels longer than 3.0 km.

3.4   CROSS-SECTION OF RAILWAY TUNNELS

One of the relevant issues for the safety of railway tunnels is the cross-sectional area of the railway galleries, especially because of the comfort conditions inside trains at high speeds.

The cross-sectional area of the railway galleries is linked to three key parameters: number of tracks per tube (because of the piston effect of the movement of the trains), speed of the trains and the permissible pressure variation inside the trains. For single-track tunnels, as there is no clash of the piston effect of the trains, a pressure variation inside the trains of 5.5 kPa is generally considered acceptable, however, in any case, it should not be higher than 10 kPa (EU 2008).

In fact, when a train enters a tunnel, the air inside is compressed, creating a compression wave that travels through the tunnel at sound speed. Subsequently, other compression and expansion waves are developed in the tunnel, mainly with high-speed trains, which are transmitted to inside the train, exposing passengers to discomfort, unless large cross-sectional areas are adopted. These aerodynamic effects have to be taken into account in the design of tunnels, especially for trains at high-speeds (Bopp 2009).

The conventional solution to reduce the pressure change inside the trains when entering a tunnel is the adoption of pressure relieves, connecting the upper part of the railway galleries, as it was adopted in the Channel Tunnel, between France and the UK, and the Groeneheart Tunnel, in The Netherlands (see, Chapter 4).

However, it is now recognised that this effect is essentially sensitive in the tunnel portals (at the entrance and exit of trains), thus the cross-sectional area of the railway galleries can be reduced if appropriate measures are adopted in these areas, which function as pressure relief, allowing to reduce the referred pressure variation inside the trains (Bopp 2009).

One solution that has been tried in single tube double-track tunnels is the local widening of the tunnel cross-section in the portal zones, as adopted in the Soumagne Tunnel, in Belgium, at the HSRL Liege-Aachen (figure 3.1) (TUC Rail 2021).

Figure 3.1.   Portal of the Soumagne Tunnel (HSRL Liege-Aachen)

In dual-tube single-track tunnels, a solution also tried is the creation of side openings along extensions of the tunnel at the portal zones, the size of which decreases from the outside into the tunnel, as adopted in the Perthus Tunnel, between Spain and France, at the HSRL Figueras-Perpignan (figure 3.2). This solution can still be improved, by the adoption of sledge edges on the portals (TUC Rail 2021).

Figure 3.2.   Portal of the Perthus Tunnel (HSRL Figueras-Perpignan)

With the adoption of these kind of measures the cross-sectional area of each railway gallery can be significantly reduced by about 15 to 20% (Bogaert 2009; Bopp 2009).

3.5 CROSS-SECTION OF ROAD TUNNELS

Regarding the vertical clearance of the road galleries, the minimum headroom above the carriageway should be at least equal to the maximum height of the vehicles that are allowed on the road, with an additional clearance to allow for vertical movements of the vehicles and drivers comfort. It varies from country to country or geographic area.

In EU countries the maximum height of the heavy goods vehicles is 4.00m (EU 1996), while in the USA, for example, higher values are allowed. If a 0.30m comfort margin is added, the minimum vertical clearance required in EU is 4.30m, although in some EU countries greater heights are required. However, if the height of vehicles allowed in the tunnel is limited, smaller values will be admitted (see, Chapter 4). In the USA the minimum height required is 4.30m in general and of 4.90m on highways.

Regarding the width of the road galleries, it should to be taken into account the number and the width of the traffic lanes (which depends on the speeds allowed in the tunnel, being in general 3.50m), as well as the inner edge and the emergency lane, when existing.

The width of the emergency lane also varies from country to country. However, it should be linked to the width of the vehicles that will park there. A generally adopted criteria is to consider the width of a passenger car, 1.75m, plus a width of 0.75m to enable drivers to get off, resulting in a total width of 2.50m. If heavy goods vehicles are allowed in the tunnel, wider emergency stop areas ("lay-bys") should be created, with a minimum width of 3.00m, at distances not exceeding 1 km along the tunnel.

The efficiency index, mentioned in Chapter 2, is a useful parameter to assess the cost/benefit ratio of each road tunnel.

3.6 EMERGENCY EXITS

Traffic (railway and road) tunnels should be provided with emergency exits, protected by escape doors, to allow tunnel users to evacuate on foot from the traffic gallery to out of the tunnel or to a safe place inside the tunnel, in an emergency situation (accident, fire, etc.) (figure 3.3) (PIARC 2011, adapted).

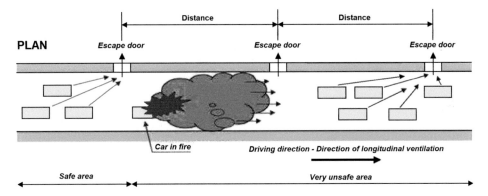

Figure 3.3. Typical escape pattern for unidirectional road tunnels with longitudinal ventilation

Emergency exits may be provided in different ways, namely:

– Cross-passages between tubes (or simply doors in the separating wall of adjacent galleries);
– Vertical access galleries connecting the traffic galleries and an underneath service gallery;
– Exits that lead into shelters, where users can remain safely during an emergency situation;
– Safety galleries disposed alongside or under the traffic galleries;
– Escape passages leading directly from an escape door to the outside.

The appropriate spacing between emergency exits is linked to several parameters (PIARC 2011), namely the types of vehicles using the tunnel, the traffic volume and the number of users, and the capabilities of the ventilation system.

As mentioned, the EU legislation (EU 2004, 2008) stipulate that the traffic galleries of railway and road tunnels should be provided with escape doors spaced 500 meters maximum. In practice quite different distance values between escape doors are used. On Table 3.1 and Table 3.2 the distances between cross-passages (escape doors) adopted in relevant railway and road dual-tube TBM tunnels, built around the world, are presented.

As it can be seen, in both cases, a wide range of distance values were adopted, between 250 meters and 500 meters in railway tunnels and between 200 meters and 800 meters in road tunnels. However, it can be shown that, for these levels of distances, the difference in the values of the distances between the escape doors is not much relevant.

Table 3.1.   Distance between escape doors of TBM dual-tube railway tunnels

	Length (km)	Distance between cross-passages (m)
Channel Tunnel (UK/France)	50.5	**375**
Guadarrama (Spain)	28.4	**250**
Koralm (Austria)	32.8	**500**
Brenner-Base (Austria/Italy)	55.0	**330**
Gothard-Base (Switzerland)	57.0	**325**

Table 3.2.   Distance between escape doors of TBM dual-tube road tunnels

	Length (km)	Distance between cross-passages (m)
Westerschelde (Netherlands)	6.6	**250**
Nanjing-Weisan (China)	3.5/4.1	**500**
Waterview (New Zealand)	4.0	**250**
M30 SBP, Madrid, Spain	3.6	**200**
Changjiang, Shanghai (China)	7.5	**800**
Sparvo (A1), Florenze (Italy)	2.6	**300**

In fact, it is known to those skilled in medical emergency, that in emergency situations the important issue is not the time to escape (for a supposed safe place inside the tunnel), but the overall time needed for rescue out of the tunnel.

The escape time of users will the sum of the walking time until the escape door used plus the time needed to cross the escape door. It is assumed that common persons walk about 50 meters per minute and the flow capacity of the escape doors is 100 persons per minute.

Let's consider an example of a train 200 meters long with 400 passengers. Two scenarios are compared: escape doors spaced 100 meters and escape doors spaced 400 meters.

In the first scenario, it is admitted that the intermediate escape door is used by 200 passengers (half of the total) and each of the adjacent escape doors are used by 100 passengers (1/4 of total each). The escape time will be of: 50/50 +200/100 = 3 minutes in the most demanded escape door and of: 50/50+100/100 = 2 minutes in the adjacent escape doors, thus the global escape time will be 3 minutes.

In the second scenario, it is admitted that 300 passengers (3/4 of the total) will use the closest escape door and 100 passengers (1/4 of the total) will use the second closest escape door. The escape time will be of: 150/50+300/100 = 6 minutes in the most demanded escape door and of: 250/50+100/100 = 6 minutes in the other escape door, thus the global escape time will be 6 minutes.

The difference of escape times between the two scenarios is, thus, of 3 minutes.

This time difference is quite negligible, when, as is known, the arrival of the rescue vehicles to the scene of an incident inside a tunnel usually takes tens of minutes; several hours, sometimes.

3.7 VENTILATION SYSTEMS

Traffic tunnels generally require the installation of ventilation systems, which serve two purposes:

- In normal operation, they should ensure sufficient air quality inside the tunnel;
- In a fire situation, they should make the environment inside the tunnel as safe as possible to users and emergency personnel, by controlling the flow of smoke in an appropriate way.

As regard railway tunnels, in normal operation, the piston effect of the moving trains is usually sufficient. However, during maintenance operations, a ventilation system is usually necessary, especially in very deep tunnels, in which temperature at the interior can be very high, over 45°C in some cases.

Regarding road tunnels, although pollutant emissions from road vehicles have decreased dramatically over the last decade, it is also generally needed to adopt a ventilation system, which will be activated, depending to the air quality inside the tunnel.

There are basically three types of ventilation in traffic tunnels: longitudinal, transverse and semi-transverse.

Longitudinal ventilation is carried out by means of jet fans installed along the traffic galleries that, by accelerating a small part of the air present in the tunnel, are able to induce, on the overall air inside the gallery, a movement in the desired direction. It is usually used in unidirectional traffic tunnels.

Transverse ventilation uses air ducts along the tunnel to introduce flows of fresh air and to extract the exhaust air from the traffic galleries through ventilation shafts arranged along the tunnel or at the tunnel portals. It is suitable in situations where longitudinal ventilation is not effective, such as the bi-directional tunnels with high traffic levels and high percentage of heavy goods vehicle traffic.

Semi-transverse ventilation uses air ducts and ventilation shafts, combined with longitudinal reversible jet-fans, to exhaust the polluted air. It is tipically used in long tunnels or bi-directional tunnels that need increased fire safety.

In fire situations, fire ventilation systems are necessary in both railway and road tunnels. Different strategies may be used, according to the type of ventilation adopted: longitudinal, transverse and semi-transverse.

The **longitudinal strategy** consists of creating a longitudinal air flow in the tunnel through jet fans, in order to push the smoke produced by a burning vehicle to the traffic direction.

However, due to the high turbulence introduced by jet fans, the action of the fans in the area of the fire must be controlled. Thus, any fans in close proximity to the fire source should not be activated and fans already in operation at those locations have to be shut down (figure 3.4) (Sturm 2017, adapted).

Jet fans should be installed at the upper part of the traffic galleries, as this is where the pollutants and smoke are most prevalent.

As users downstream of the fire may be affected by the toxic gases and reduced visibility, this strategy should only be used in unidirectional traffic tunnels.

The **transverse strategy** provide the possibility of extracting smoke close to the fire location. It takes advantage of the buoyancy of fire smoke, that is, smoke tends to concentrate

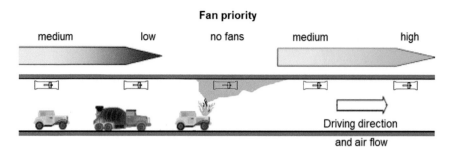

Figure 3.4.   Fan activation strategy in road unidirectional tunnels with longitudinal fire ventilation

in the upper part of the tunnel space, from where it can be mechanically extracted, through dampers regularly arranged along the tunnel.

The system is designed so as to preserve a fresh air layer in the lower part of the tunnel (correct visibility, low toxicity), in order to facilitate self-evacuation of users. The effectiveness of this strategy depends on the possibility of confining smoke within a short region (control of air/smoke flow). Thus, longitudinal ducts are arranged in the tunnel (at the top and/or at the bottom) and ventilation shafts with axial fans are installed along the tunnel or at the tunnel portals, to inject fresh air and exhaust smoke throughout the tunnel (figure 3.5) (Sturm 2017, adapted).

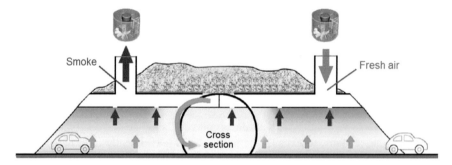

Figure 3.5.   Transverse fire ventilation in road tunnels

This strategy is applicable to any tunnel, however, it is more expensive and complicated to operate than longitudinal ventilation systems.

In the **semi-transverse strategy**, the exhaust system with a longitudinal duct and ventilation shafts is complemented with longitudinal jet fans installed along the tunnel, to force the movement of the smoke towards the extraction dampers. Thus, longitudinal reversible fans are regularly arranged along the tunnel, to force the movement of the air from the tunnel portals towards the extraction site, which allows to minimize the power required for the transverse ventilation system (figure 3.6) (Sturm 2017, adapted).

In this case, the jet fans are usually used as the ventilation system in normal operation.

The design of appropriate ventilation control scenarios, for each fire situation in the tunnel is a very important part of the process. These scenarios can be simple, especially when the longitudinal strategy is applied, or involve a large number of measurements and ventilation devices in complex transversely-ventilated tunnels.

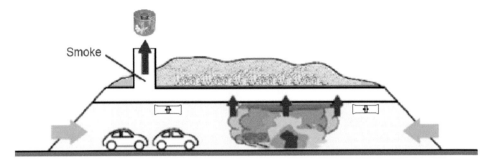

Figure 3.6. Semi-transverse fire ventilation in road tunnels

With regard to railway tunnels, medium-sized tunnels are usually provided with longitudinal ventilation systems, however being used mostly during maintenance operations or in fire situations. In fact, in normal operation, the piston effect of the trains in circulation is usually sufficient to ventilate the tunnel.

However, in long railway tunnels, in addition to longitudinal ventilation, transverse ventilation systems, with ventilation shafts and powerful ventilators, are sometimes installed at the tunnel portals or in the so-called "rescue stations", areas in the tunnel prepared for the stop of trains in emergency situations.

With regard to road tunnels, as mentioned, a fire ventilation system is usually necessary along the entire length of the traffic galleries. Longitudinal or semi-transverse ventilation systems are usually adopted. Key elements for the choice are: length, unidirectional or bidirectional traffic, gradient, cross-section, peak traffic flow rates or fire size. In very short flat tunnels (less than 0.5 km), natural ventilation can be sufficient in normal operation.

The optimization of the ventilation control during normal operation is a very important task in order to reduce energy consumption, since this consumption represents significant part of the operational cost of the tunnel.

In both railway and road tunnels, other parts than the traffic galleries can require own-ventilation, such as the emergency exits and the evacuation galleries.

The operation of the fire ventilation system must be controlled in connection with the operation of the Fixed Fire Fighting System (FFFS) of the tunnel, if installed.

In Chapter 4, noteworthy examples of railway and road tunnels are presented, with a description of the main aspects of the respective ventilation systems.

3.8 FIRE FIGHTING SYSTEMS

3.8.1 General

The primary objective of the fire fighting system in a tunnel is to provide the means to fight a fire within the tunnel with minimum impact on the users, the emergency personnel and the tunnel structure.

The systems critical to the ability to fight a fire within a tunnel include: detection, emergency ventilation, water supply and distribution, and FFFS ("Fixed Fire Fighting System"). In road tunnels it usually also include alarm, radio communications, emergency telephone, close television circuit, loudspeakers, and portable fire extinguishers. These systems must be planned and installed in a careful integrated manner.

A water supply system, including water mains and fire lines, is required to provide water for fire fighting within the tunnel (through hydrants or hose valves) and to provide water for the WFS ("Water Based Fixed Fire Fighting System"), if installed in the tunnel.

The source of water can be from a water distribution system or from a water tank with enough capacity. The required system pressure must match the requirements of the responding fire brigade and of the FFFS.

Fire hydrants (hose valves) are required to provide points of connection for the fire fighters to attach fire hoses and have access to the water supply. They should be installed at regular intervals within the tunnel.

In road tunnels, portable extinguishers are also provided at regular intervals, to allow motorists to fight low size fires, prior to the arrival of the fire brigade.

The strategy for fire attack must be established on the Safety Plan of the tunnel. For example, in dual-tube tunnels, it needs to be clear whether fire trucks will use only one or both portals in the incident gallery, or the four portals in both galleries.

3.8.2   Fixed Fire Fighting Systems

In a rapidly developing fire, smoke can quickly compromise the ability of users to rescue themselves, while a rapid rise in temperatures, usually reaching more than 1000°C, can destroy safety systems and make problematic the action of fire fighters in attacking the fire. In addition, the structure can be deeply damaged, requiring long periods of service interruption for repair works.

An FFFS ("Fixed Fire Fighting System") has the potential to reduce the rates of fire growth and spread, thereby assisting the emergency services during the self-rescue and assisted-rescue phases of a fire. Other potential benefits of an FFFS are the protection of the tunnel assets from fire damage and to avoid or reduce the interruptions that may occur on the service provided the tunnel, while is being repaired.

Rapid and accurate incident detection and FFFS response are essential components to obtain the best performance. The FFFS must be designed in coordination with the design of the ventilation system of the tunnel. With the installation of an FFFS, the sophisticated transverse or semi-transverse ventilation systems can, in principle, be replaced by simple longitudinal ventilation with jet fans, which are simpler and much less expensive.

The WFSs ("Water Based Fixed Fire Fighting Systems") are by far the most efficient FFFSs in tunnels. However, other water-based systems, including foam systems, are also available. The selection of the appropriate FFFS should be based on cost-benefit analysis.

WFS use water in the form of droplets. Depending on the type of system, the average droplet size can vary from relatively large droplets, called Deluge Sprinkler Systems, to small droplets, called High Pressure Water Mist Systems. Systems can be of Low Pressure (less than 10 bar) to High-pressure (up to 140 bar), with the later having smaller droplet sizes. Water Mist Systems consume much lesser water than sprinkler systems.

As a result of the extinguishing water being split up into droplets, a reaction surface is created via which the heat from the fire is absorbed, dramatically lowering the temperature at the fire site. Figure 3.7 shows the temperature evolution of a fire when a WFS is activated 15 minutes after the start of the fire (Feltmann 2015, adapted).

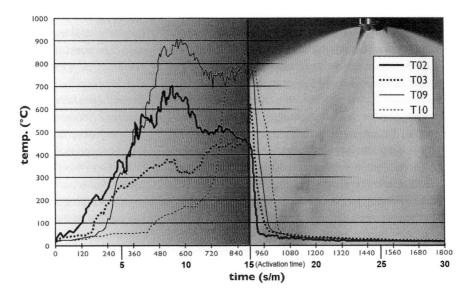

Figure 3.7. Temperature evolutiom of a fire with the activation of a WFS

The smaller the mean droplet size, the more efficient the cooling effect (on the air and gases around the fire). As at high temperatures, during water evaporation its volume increases more than 1000 times, there is a reduction of the oxygen content in the air, and thus the fire will be suffocated, or at least, will be suppressed.

In fact, WFSs are not meant to extinguish fires, but to suppress them, that is, to control them and mitigate their effects. The complete extinguishing of the fire is the responsibility of the fire brigade, the point is that their task will be much easier, since the temperatures will be much lower, thus they can get much closer to the source of fire (SOLIT 2012).

To ensure fast, effective and efficient fire suppression, groups of nozzles forming sections shal be activated simultaneously in the fire site and in the area up and downstream of the fire. All sections will be connected by section valves to a main water supply line to the pump unit. Figure 3.8 shows a WFS in action in a road tunnel (Fogtec 2019).

The water supply shall be sufficient such that the WFS provides the required flow water rate for a minimum of 30 minutes, or for a period of time that is double the time required for the fire fighters to reach the fire.

WFSs are "area protection systems", being installed at the upper part of the side walls (at least at a height of maximum vehicle height), pointing downwards and covering the entire traffic width. Thus, the entire tunnel or just "protected areas" should be covered with nozzles, which will be grouped into sections, with a minimum recommended length of 30 meters (UPTUN 2007).

With this effect of temperature reduction inside the tunnel, the use of WFS systems make also possible to reduce the requirements on concrete protection and the fire rating of mechanical and electrical equipment.

On Chapter 4, noteworthy examples of railway and road tunnels are presented, with the description of the main aspects of the respective fire fighting system.

Figure 3.8.   WFS in action in a road tunnel

3.9   SAFETY IN OPERATION

Safety in operation of the tunnel, also called the safety concept of the tunnel, is meant the set of procedures provided in the tunnel for the evacuation of users to out of the tunnel in the event of an accident or fire inside the tunnel. They must be established in the Safety Plan of the tunnel.

With regard to road tunnels, as mentioned, tunnels over 0.5 km in length must be provided with emergency exits (escape doors) spaced 500m at the most (at least in the EU). In the event of an incident inside one of the road galleries of the tunnel, drivers will be advised to stop and to leave their vehicles (as well as the other passengers of those vehicles) and to walk to the nearest escape door, in order to take an emergency exit to outside, or the corresponding cross-passage that take to the other traffic gallery, from where they will be rescued later by means of rescue vehicles.

The strategy for the entry of the rescue vehicles into the tunnel, namely, if from one or from both portals of the non-incident road gallery, must be previously established in the Safety Plan of the tunnel.

With regard to railway tunnels, procedures will in principle depend on the tunnel length.

In medium-size tunnels (between 1 km and 20 km in length), as a rule, when a fire be detected on a train, the driver should try to take the train out of the tunnel. Anyway, if the train stops in the meantime, users and staff will be asked to leave the train and find a safe place (usually, the non-incident gallery), using the emergency exits (escape doors), which should also be spaced 500m at the most (at least in EU countries).

In long railway tunnels (over 20 km long) procedures will be more complex; different strategies are being adopted.

If the tunnel is provided with a service gallery, connected to the railway galleries by cross-passages, users are asked to leave the incident gallery and to go to the service gallery, from where they will later be rescued by emergency vehicles, which circulate in this gallery.

In case the tunnel is not provided with a service gallery, other solutions are being adopted.

One solution is to provide the tunnel with a "safe area" (an "emergency room") halfway, where users are able to survive for several hours, waiting for a rescue train that will access through the non-incident gallery. However, in order the solution be effective, the distances that users will take to reach the "emergency room" should not be too long.

Another solution is the creation of so-called "rescue stations" inside the tunnel, where users have special rescue conditions, namely these areas are usually equipped with powerful ventilation shafts that provide fresh air to users and extract polluted air from the tunnel.

Thus, when a fire be detected, the driver will try to stop in one of those "rescue stations", where passengers leave the incident train to the other railway gallery, from where they will be later rescued by a rescue train. Here too, for the solution to be effective the distances between "rescue stations" should not be too long, although distances greater than 20 km have already been adopted.

In the case the incident train stops outside these "rescue stations", passengers will still be asked to leave the train and pass (through the cross-passages) to the other railway gallery, from where they will later be rescued by the rescue train, however without the comfort conditions offered in the "rescue stations".

On Chapter 4, noteworthy examples of railway and road tunnels are presented, with the description of the main measures planned to ensure safety during the operation.

3.10 CONCLUSIONS

The safety of traffic (railway and road) tunnels has received great attention in recent time, namely at EU level, which issued two reference documents: EU Decision 163CE/2008: "Safety in Railway Tunnels",mandatory for railway tunnels with more than 1.0 km; and Directive 2004/54/CE: "Safety in Roadway Tunnels", which applies to road tunnels over 0.5 km that are part of the Trans-European Road Network.

Both documents present recommendations and measures to be adopted in order to reduce the risk during operation of these tunnels.

As for the cross section of the railway tunnels, it is linked to the comfort conditions inside the trains, at the entrance and exit of the trains in the tunnel portals. Although not explicitly foreseen on the EU Decision, the adoption of appropriate pressure relieves in these areas allows to reduce the cross-sectional area of the referred railway galleries.

The cross section of road tunnels is linked to the number and width of the traffic lanes in the tunnel, and the height of the vehicles authorized to use the tunnel. The efficiency index of the tunnel cross section is a useful indicator to assess the cost/benefit ratio of road tunnel solutions.

Traffic tunnels (railway and road) must be provided with emergency exits, protected by escape doors, to allow the evacuation of users to out of the tunnel (or to a safe place inside the tunnel) in an emergency situation (accident or fire). The aforementioned EU documents require that emergency exits (escape doors) have a maximum spacing of 500 meters. In practice, quite different values are adopted. However, as it was shown, for this level of distances, the difference in distances between the escape doors is not much relevant.

Ventilation systems are also required in railway and road tunnels, particularly to act in fire situations. Different strategies can be adopted; longitudinal ventilation (the simplest), and transverse and semi-transverse ventilation systems, which are more complex and more expensive. The operation of the fire ventilation system must be controlled in connection with the operation of the Fixed Fire Fighting System (FFFS) of the tunnel, if installed.

Traffic tunnels also require a fire fighting system, being, traditionally, based on conventional fire trucks. Modern FFFSs ("Fixed Fire Fighting Systems"), in particular WFSs ("Water Based Fixed Fire Fighting Systems"), although more expensive, are more reliable, allowing for a drastic reduction of the temperature at the site of the fire, greatly facilitating the task of firefighters and significantly reducing the effects on the structure and equipment. Thus, considering the tunnel life-cycle, the installation of an FFFS will, in principle, be more cost-effective.

Tunnels must have be provided with a Safety Plan, which establish the procedures to ensure safety in operation (safety concept), that is, the set of procedures for the evacuation of users in the event of an accident or fire inside the tunnel. It is assumed that emergency exits (escape doors) will provide access to outside the tunnel or the non-incident gallery (in double-gallery tunnels). Different strategies may be adopted, especially in the case of long railway tunnels, although their reliability is not fully guaranteed.

REFERENCES

Bogaert P (2009), *Current and End Section Design Requirements for High-Speed Railway Tunnels.* Workshop Tunnels for High Speed Railway, Porto, Portugal, October 2009

Bopp R, Hagenah B (2009), *Aerodynamics, Ventilation and Tunnel Safety for High Speed Rail Tunnels*. Workshop Tunnels for High Speed Railway, Porto, Portugal, October 2009

EU (1996), *Weights and Dimensions.* EU Directive 1996/53CE, July 1996

EU (2004), *Safety in Roadway Tunnels* EU Directive 2004/54CE, April 2004

EU (2008), *Safety in Railway Tunnels.* EU Decision 2008/163CE, January 2008

Feltmann A et al. (2015), *Upgrading Tunnels through Active Fire-Fighting Systems. Part 1- Technical and Economic Marginal Conditions*. Tunnel N. 8, December 2015

Fogtec (2019), *The Smarter Way of Firefighting.* Fogtec, 2019

PIARC (2011), *PIARC Road Tunnels Manual.* PIARC, 2011

SOLIT (2012), *Engineering Guidance for a Comprehensive Evaluation of Tunnels with Fixed Fire Fighting Systems Using the Example of Water Based Fixed Fire Fighting Systems.* SOLIT (Safety of Life in Tunnels), Scientific Final Report of the SOLIT Research Project Prepared by the SOLIT Research Consortium, 2012

Sturm P et al. (2017), *On the Problem of Ventilation Control in Case of a Tunnel Fire Event.* Elsevier. Case Studies in Fire Safety, Volume 7, May 2017

Tielkes T (2006), *Aerodynamic Aspects of Maglev Systems*. 19th Conference on Magnetically Levitated Systems and Linear Drives, Dresden, Germany, September 2006

UIC (2003), *UIC Code 779-9: Safety in Railway Tunnels.* UIC, 2003

UPTUN (2007), *Engineering Guidance for Water Based Fire Fighting Systems for the Fire Protection of Tunnels and Subsurface Facilities.* UPTUN, WP2 of the Research Project UPTUN, R251, August 2007

Chapter 4

Current concepts in TBM traffic tunnels and noteworthy examples

Abstract

This chapter presents the most advanced current concepts for TBM traffic tunnels: the three-tube tunnel concept, the dual-tube tunnel concept and the mono-tube dual-gallery tunnel concept for railway tunnels; and the dual-tube tunnel concept and the mono-tube double-deck tunnel concept for road tunnels.

Noteworthy examples of railway tunnels (Channel Tunnel, Gotthard-Base, Groenehart tunnel, etc.) and of road tunnels (Duplex Tunnel, SR99 Tunnel, etc.), built around the world in the recent decades, based on these concepts, are presented, focusing on aspects, such as geometry, construction methods and technical systems (namely, ventilation and fire fighting), and safety in operation. The main achievements and shortcomings in each case are also mentioned.

4.1 INTRODUCTION

For more than a century, long railway tunnels, with lengths over 10 km, have been steadily built, some of the oldest still in operation today. The conventional solution for a long time, was the single tube twin-track tunnels. As the safety requirements imposed on railway tunnels have however increased dramatically, especially in the past two or three decades, this solution has been severely limited, unless drastic operational restrictions are applied. In fact, modern railway tunnels are highly complex technical systems with high demands. Thus, to earn public's trust, the railway infrastructure of today must, on the one hand, provide economic exploitation, and, on the other, meet demanding safety requirements.

With regard to road tunnels, although with some delay, a similar process took place, although perhaps without reaching a similar scale. For a long time, road tunnels were also made of single tubes enclosing the two traffic directions. However, with the new safety requirements, especially from the beginning of the 2000s, this is no longer possible, unless complex safety measures are adopted and strong limitations are imposed on traffic inside the tunnel, namely the transport of heavy goods and the speed of the vehicles. It has to be noted that tunnels in roadway networks are today an essential component for global mobility, in both the urban areas and the major traffic axis.

Consequently, in order to face the new challenges, more advanced concepts for railway and road tunnels have been developed and are being adopted, particularly in those built with the TBM technique.

DOI 10.1201/9781003144861-5

4.2 TBM RAILWAY TUNNELS

4.2.1 Current concepts in TBM railway tunnels

Until the 1990s, railway tunnels built using the TBM technique (and the others) were generally formed by single tubes with two tracks placed side by side.

With the construction of the challenging Channel Tunnel, between France and UK (United Kingdom), safety requirements in railway tunnels were thoroughly analysed, which gave rise to the development of new concepts with a view to improving safety.

A recognized innovative concept was thus developed and adopted: the **three-tube tunnel concept**, in which the tunnel is formed by three TBM tubes, two single-track tubes (each for a traffic direction) and a service tube placed between the other two, for safety and technical purposes: evacuation of users, maintenance services, installation of pipes, etc., the three connected by cross-passages regularly arranged along the tunnel (figure 4.1a).

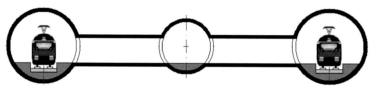

a) Three-tube tunnel with two single-track tubes

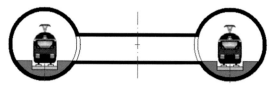

b) Dual-tube tunnel with single-track tubes

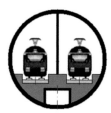

c) Mono-tube tunnel with two single-track galleries

Figure 4.1.   Current concepts in TBM railway tunnels

With the experience gained some of the innovations implemented were questioned, thus a simplified concept was in the meantime developed, the **dual-tube tunnel concept**, in which the tunnel is formed by two single-track TBM tubes connected by cross-passages, which became a kind of reference solution for the new railway tunnels, particularly in HSRL (High-Speed Railway Lines) (figure 4.1b).

However, a very reliable and even more simplified concept was also developed, the **mono-tube dual-gallery tunnel concept,** in which the tunnel consist of a single TBM tube, housing two isolated single track railway galleries, placed next to each other (figure 4.1c).

Below are some noteworthy examples of railway tunnels built on the basis each of these concepts, focusing on aspects such as geometry, construction methods and technical systems, namely ventilation and fire fighting, and safety in operation.

4.2.2   Three-tube tunnels with two single-track tubes

So far, there is only an example of a three-tube railway tunnel with two single-track tubes built, the Channel Tunnel, between UK and France, under the English Channel, which is presented below. Another example in which this concept is foreseen to be used, the Strait of Gibraltar tunnel, is still in the planning stage.

Channel Tunnel, UK/France

The Channel Tunnel, also known simply as Eurotunnel, links Folkstone in Kent (UK), to Coquelles in Pas-de-Calais (France), under the British Channel (La Manche, for the French).

The tunnel is 50.5 km long and goes down 100 meters under the sea level, being the section under the sea, 37 km long (figure 4.2) Getlink 2021), the longest under water tunnel in the world. The journey time for the crossing is 27 minutes (120 km/h maximum speed).

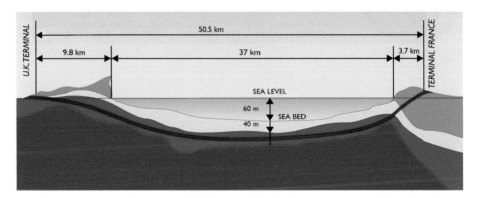

Figure 4.2.   Channel Tunnel. Longitudinal section

Ideas for a fixed link crossing in the English Channel date back to Albert Mathieu in 1802, involving horse-drawn carts and an island built in the middle of the Channel (Dennis 2019).

However, the project was not well regarded on the English side, with British politicians and the press finding that national security would be compromised. As mentioned in Chapter 2, in the early 1880s excavations were started and more than 1 km of tunnel was built on both sides of the Channel, using primitive TBMs. However, under pressure from the English side, in 1883 the project was abandoned.

With the entry of UK in the European Communities (predecessor of the European Union, EU) in the early 1970s, construction of the tunnel, financed by the French and UK governments, started on both sides of the Channel, but the project was cancelled by the UK government, invoking financial concerns. At the time, proposals were also made for a bridge crossing, but were not followed.

Finally, in 1985, the UK and French governments agreed to invite submissions for a fixed link. Eurotunnel, a group of ten construction companies and five banks, was awarded the project, based on the 1974 attempt. It was a BOOT (Build-Own-Operate-Transfer) project with a concession for 65 years (55 years, initially).

Tunneling started in 1988 and the tunnel began operating in 1994 (6 years of works). At current prices, cost was about EUR 13 000 million (EUR 257 million per km).

The construction of the Channel Tunnel represents the affirmation of the TBM technique in the construction of tunnels worldwide. In 1996 the American Society of Civil Engineers (ASCE) classified the tunnel as one of the "Seven Wonders of the Modern World".

The tunnel consist of three TBM tubes, two single track tubes for mixed railway traffic (Eurotunnel shuttles, Eurostar and national freight trains) and a service tube placed in the middle of the other two (figure 4.3) (Getlink 2021, adapted).

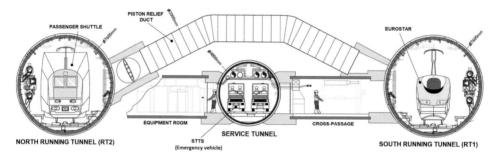

Figure 4.3.   Channel Tunnel. Cross-section

The railway tubes are connected to the service tube by cross-passages placed every 375 meters. They are also connected to each other by upper piston relief ducts 2.0m diameter every 250 meters.

The railway tubes are 7.6m inner diameter and 30m apart. Each one has a single track, and two walkways (one for maintenance purposes and the other for use in the event of an emergency evacuation). The walkways are also designed as anti-derailment devices.

The service tube is 4.8m inner diameter and lies between the two rail tubes. Effectively, it is a road tunnel used by electric and diesel-powered vehicles.

The tunnel was bored essentially through a layer of chalk marl, a waterproof ground (to minimize the risk of water entrance), easy to excavate and very resistant. Eleven TBMs were used, 6 on the English side and 5 on the French side (one of the TBMs bored the two main tubes inland), that start excavation from shafts built close to both margins of the channel (Robbins 2021). The service tunnel was built ahead of the others, functioning as pilot tunnel.

On the English side, Double Shield TBMs were used, with excavation diameters of 8.36m in the case of the railway tubes.

On the French side, due to the higher permeability of the ground, TBMs were of the EPB type, for 10 bars pressure and 8.80m excavation diameter, although being used mostly in open mode.

The track in each rail tunnel has continuously welded rails laid on precast concrete supports, embedded in the concrete track bed.

The railway track system is quite complex. In normal operations trains use the South tunnel (RT1) in the France – UK direction, and the North tunnel (RT2) when travelling from the UK to France. Two undersea crossovers allow trains to pass from one rail tube to the other during maintenance periods, or when one of the tunnels becomes inoperative (figure 4.4) (Getlink 2021, adapted).

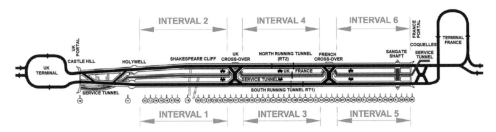

Figure 4.4.   Channel Tunnel. Diagram of the railway connections

The tunnel is provided with a longitudinal ventilation system with ventilation buildings in Shakespeare Clift (UK) and Sangatte (France), supplemented by ventilators on either ends of the tunnel. In the event of a fire, ventilation is used to move smoke in the traffic direction in the railway tubes to give passengers clean air. Air pressure is higher in the service tube to prevent the ingress of smoke in the event of a fire in one of the railway tubes.

As mentioned, to contrary the piston-effect pressure created by trains at high speeds, piston relief ducts of 2m diameter were built every 250m. However, this solution led to the creation of unacceptable lateral forces on the trains, so it was decided a reduction in the speed of the trains and the installation of restrictors in these ducts.

The initial fire fighting system was based on attack by fire brigades, moving in dedicated vehicles through the service tunnel. However, this system has not proved to be very effective. In fact, in 1996 a heavy goods vehicle shuttle wagon fire caused severely damage on the structure and restricted use of the tunnel for five months. A less severe incident occurred in 2006 when a lorry aboard a shuttle train caught fire, causing also disruption of the traffic in both directions during several days.

Thus, in 2012, a new fire extinguishing system, an WFS ("Water Based Fire Fighting System") was installed in the tunnel. It is formed by four "safe stations" 810 meters long (spaced about 10 km), each divided into 27 sections 30 meters long each, equipped with high-pressure water mist nozzles (Reynolds 2011). However, the installation of this new system was quite expensive, since it required a very complicated adaptation works.

The service tube, in addition to allowing access to maintenance teams, allows access to rescue teams and serves as a safe haven if passengers need to be evacuated in an incident situation. In the event of an incident that obliges the train to stop, users will leave the incident train and walk to the service tunnel, to be later rescued by special emergency vehicles.

A fleet of dozens of emergency wheeled vehicles, the STTS (Service Tunnel Transportation System), were specifically designed for travel inside the service tunnel (figure 4.5) (Holt 2017).

Figure 4.5.   Channel Tunnel. Emergency vehicle (STTS)

The entire transport system is controlled from the RCC (Rail Control Centre), which functions in two centers, one on each terminal (see, figure 4.4).

Remarks

The Channel Tunnel, a railway tunnel built on the basis of the three-tube concept, is an impressive project, however, the only so far built on the basis of this concept, although it was considered at the time the most promising for making long tunnels.

With more than 25 years in operation it can be said that it fulfilled its mission, but revealing some gaps. To be mentioned the very high cost (about EUR 257 million per km) and weaknesses in the technical systems, in particular on the fire fighting system, which lead to its replacement by a new one, an FFFS ("Fixed Fire Fighting System") however becoming quite expensive, as very complicated adaptation work was required. Also the pressure relief system of the railway galleries as well the rescue system of users (safety in operation) in the event of fire or accident are not completely satisfactory.

4.2.3 Dual-tube tunnels with single-track tubes

A large number of dual-tube single-track railway tunnels have been built. Due to their importance and specificity, the following cases are presented: the Guadarrama Tunnel, in Spain, and the Gotthard-Base Tunnel, in Switzerland.

Guadarrama Tunnel, Spain

Spain has the longest high-speed railway network in Europe and the second highest in the world, after China. The country's total high-speed network is over 3000 km, with five lines dedicated to passengers traffic, which establish high-speed links between the capital, Madrid, and the largest cities of the continental Spain (ADIF 2021a).

The Guadarrama Tunnel is located along the HSRL (High-Speed Railway Line) Madrid – Valladolid, in the Sierra de Guadarrama, 45 km North of Madrid. It runs between Miraflores de la Sierra and Segovia, allowing the Madrid-Valladolid stretch, 175 km long, be covered in 55 minutes. The project was carried out under the responsibility of ADIF, the Spanish authority for railway infrastructure.

The tunnel is a dual-tube tunnel, 28.4 km long, the fourth longest in Europe and the fifth in the world, allowing train speeds of 300 km/h (figure 4.6) (ADIF 2021a). The budget granted was EUR 1220 million (EUR 43 million per km).

Figure 4.6.   Guadarrama Tunnel. A view of the twin-tubes at the South portal

The two tubes are 30 meters apart and are connected by cross-passages located every 250 meters. The inner diameter of the tubes is 8.50m, using precast segments 0.32m thick 0.18m gap to be injected, thus, excavated diameters of 9.50 meters.

The construction works began on September 2002, being completed in approximately 32 months. The tunnel was opened to traffic in December 2007.

The geological profile shows five major sections depending on the character of the principal units traversed by the route. There are granites and gneiss with a high degree of consistency, hardness and abrasiveness (figure 4.7) (ADIF 2021b, adapted).

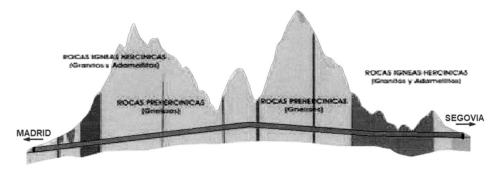

Figure 4.7.   Guadarrama Tunnel. Longitudinal section

The construction of the tunnel had to face significant constraints, since the construction process was developed based on criteria of meeting the completion deadline and having the minimum possible environmental impact.

In fact, the Guadarrama massif is classified an especially protected area, therefore measures were taken to cause the least environmental impact on their surroundings and the environmental integration of the project, protecting the hydrological system, fauna and flora.

In addition, following the Environmental Impact Statement, the excavated material should be reused, as much as possible, to manufacture concrete for the precast segments.

The use of intermediate stages could therefore not be used in the construction of the tunnel, so four TBMs were used, two to bore each of the tubes, two supplied by Herrenknecht AG and the other two supplied by Wirth.

The TBMs were designed specifically for extra hard rock, adapted for the characteristics of the route and the geology of the massif. TBMs are of the Double shield type, which offer high levels of safety and excavation rates in that type of ground. In those 32 months, the four machines excavated 4 million cubic meters of rock.

Each of the TBMs, two working from the South portal (Madrid side) and the other two from the North portal (Segovia side), advanced at an average pace of 16 meters per day. In some periods, close to 1,000 meters of tunnel were excavated in one month by a machine.

Each of the tunnel rings is made up of 7 precast segments 1.60m wide, leading to a total of about 250 thousand units. These were manufactured at the worksite, each of the four TBMs assured of its supply by its own manufacturing plant.

Ventilation buildings were built, located in the portal areas, each equipped with reversible fans, to, in case of fire, ventilate the incident gallery and pressurize the non- incident gallery (figure 4.8) (ADIF 2021c, adapted).

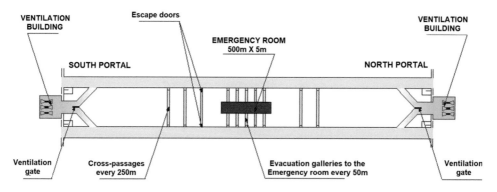

Figure 4.8. Guadarrama Tunnel. Scheme of the ventilation and safety systems

Fire fighting is carried out by conventional means, through fire brigades that enter the tunnel galleries through the portals. Fire detectors and hydrants are arranged, at regular intervals, along the tunnel galleries.

In the event of a fire, if a train stops, traffic will be stopped in both galleries. Then, passengers will leave the incident train to the non-incident gallery, through the cross-passages, where they will wait for a rescue train that will circulate later in that gallery.

At the middle of the length of the tunnel, placed between the two rail galleries, there is an emergency room, 500m long and 5m wide, which is equipped with its own ventilation system that guarantees fresh air to users for 48 hours in the event of a fire.

The tunnel is controlled by a Control Center that monitors all safety equipment.

Gotthard-Base Tunnel, Switzerland

The Gottthard-Base Tunnel, in southern Switzerland, is a key component of the NRLA (New Rail Link through the Alps), under the Gotthard and Lötschberg mountains, a high-speed rail route for mixed traffic, in the N-S direction, at altitude 500–600. It is also part of the Rhine-Alps (Rotterdam-Genoa) corridor of the TNT-T (Trans-European Transport Network) (Anagnostou 2013; Fabbri 2019).

The tunnel connects Erstfeld, in the north, to Bodio, near Lugano, in the south. It is 57.1 km long, the longest railway tunnel in the world. Design speeds are 250 km/h for passenger trains and 160 km/h for freight trains. About 150 trains use the tunnel daily, of which two thirds are freight trains, but it is foreseen that they will be 50% more in the near future.

After approval in a popular referendum in 1992, excavation of the tunnel began in 1999, having been concluded in 2011 and the tunnel started operating in December 2016. The cost was equivalent to about EUR 8500 million (EUR 149 million per km).

The project was carried out under the responsibility of AlpTransit Gotthard Ltd (ATG), a subsidiary of the Swiss Federal Railways (SBB-CFF-FFS). In 2018 it was awarded the European Railway Award of the year.

The tunnel consists of two parallel ballastless railway tubes of 8.40m inner diameter and 40m of distance between axis, connected by cross-passages every 325 meters (figure 4.9) (Fabbri 2019, adapted).

The tunnel is provided with two MFSs (Multi-functional Stations), 600 meters long that include emergency stop stations in each tube, in the event of an incident, and train diversion via cross-over to the other tube, as well as access galleries and ventilation shafts. The distances between the MFSs and from these to the portals are 23 km, 18 km and 16 km, respectively, from North to South.

The construction of the tunnel has to face significant constraints, as the high variability of the characteristics of the rock to be bored, and the great thickness of the overburden that reaches 2450 meters (the deepest tunnel in the world), provoking temperatures inside the tunnel over 45°C.

Due to its great length, the construction of the tunnel was divided into five sections: Erstfeld, Amsteg, Sedrun, Faido and Bodio (figure 4.10) (Wikipedia 2021a, adapted), which were carried out by five independent contractors, starting in different access points along the tunnel, however not obliged to use the same excavation methods.

The tunnel was mostly driven (80%) by four identical TBMs of the Gripper (Open) type, supplied by Herrenknecht AG, in a total length of 45 km (Herrenknecht 2016), while in about 12 km of total length, conventional driven methods (mostly, drill-and-blast) were used.

The excavated diameter of the tunnel tubes varies according to the rock conditions, being 9.55m as standard in the TBM sections, with 0.35m thick precast segments (figure 4.11a). In squeezing (overstressed) rock areas the lining consists of an umbrella waterproofing with an unreinforced shell 0.60m to 1.20m thickness (figure 4.11b) (Anagnostou 2013, adapted).

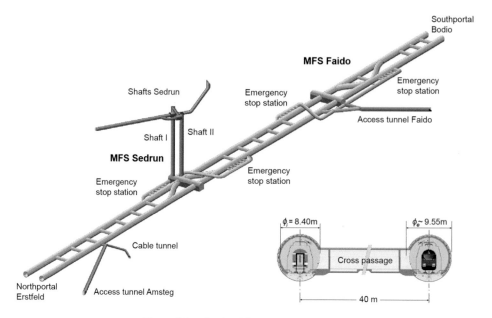

Figure 4.9.   Gotthard-Base Tunnel. Scheme

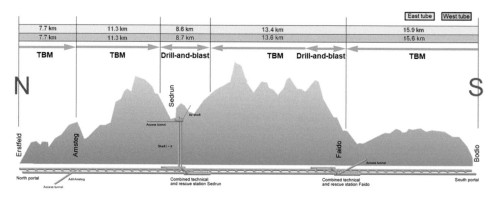

Figure 4.10.   Gotthard-Base Tunnel. Excavation methods

The two tubes of the northern sections, Erstfeld and Amsteg, totalling 19 km each, were driven by TBMs, in both cases from north to south. However, due to delays in the contract process approval, the Amsteg section was driven first, starting from an access gallery purpose built. After reaching the boundary of the Sedrun MFS, both TBMs were dismantled and transported to Erstfeld, via de access tunnel Amsteg. Afterwards, with the same TBMs the Erstfeld section was driven.

The Sedrun section also included excavation of the Sedrun MFS. It was built using conventional tunnelling methods: drill-and-blast, as well as mechanical excavation, using a hydraulic hammer.

The access to the Sedrun MFS site was made through a level access tunnel, 1 km long.

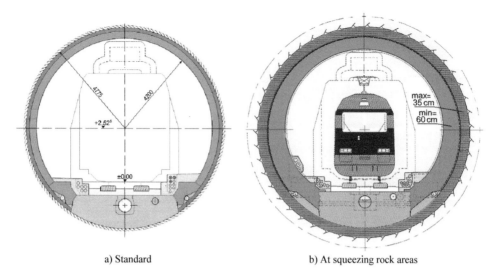

a) Standard   b) At squeezing rock areas

Figure 4.11. Gotthard-Base Tunnel. Cross-section of the tunnel tubes

The Faido section, which also includes the Faido MFS, was excavated by drill-and-blast. Because of the fault conditions of the rock encountered at the initial stage of excavation, the cross-over zone and the emergency stop station of the west tube were moved by around 600m to south (see, figure 4.9). Access to the site was provided by an access gallery 2.7 km long.

The Bodio section was excavated by two TBMs, starting at the South portal. Although encountering some horizontal faults with squeezing effects, the excavation was successfully overcome. After these breakthroughs, the cutter heads of the two TBMs were dismantled and the machines were transported to the north end of the Faido's MFS, where they were reassembled and the excavation restarted to reach the boundary of the Sedrun section.

In total more than 13 million cubic meters of rock were excavated.

The safety systems of the tunnel are quite complex and expensive, having been revised and enhanced after fatal fires in Swiss tunnels during the period of its construction (Ehrbar 2016; Kuhn 2016).

Regarding ventilation, height large axial fans (the world's most powerful), developed and built especially for this application, were installed in very deep ventilation shafts created at the ends of each MFS, to exchange the air inside the tunnel. In addition, 6 jet fans were also installed in the portal areas of each tube, in groups of two, spaced about 80m.

With regard to safety in operation, rescue of users in the tunnel take place exclusively via rail. There are two intervention centres (Erstfeld on north and Biasca on south), in which emergency trains are ready to intervene (Luetsche 2017).

If a train reports a fire, via sensors that are mounted on the train and in the tunnel, it is routed to the nearest MFS and instructed to make an emergency stop. Thus, four fans extract smoke from the emergency station while the other four inject fresh air. Fire and train localization determine the direction of smoke control.

The attack on fire is carried out by conventional means, through fire brigades that access the site by means of an emergency train that enters the tunnel via reverse mode. Water points along the MFSs provide pressurized water to fire fighters. After the fire is extinguished, the incident train will be towed out of the tunnel by the emergency train (figure 4.12) (Luetsche 2017, adapted).

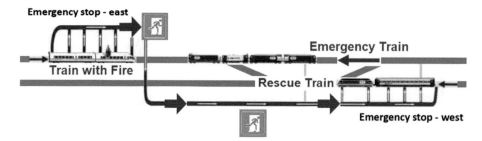

Figure 4.12.   Gotthard base Tunnel. Safety in operation concept

Evacuation of passengers is the last step. At the emergency station, if possible, or via cross-passages, passengers rescue themselves and help people with disabilities to escape to the non-incident tube, walking through a complex system of galleries, where a rescue train will afterwards take them out of the tunnel.

Remarks
The two dual-tube single-track railway tunnels presented above, Guadarrama and Gotthard-Base are two very impressive projects, some of the longest tunnels in the world, the first for high-speed passenger's traffic, and the second for mixed traffic (freight and passengers).

Although the global lay-outs are similar, they differ significantly.

As for the cost, although the ground conditions are not very different, the cost of the Guadarrama, was only EUR 43 million per km, while that of the Gotthard-Base was very high, about EUR 149 million per km (more than three times higher).

As for ventilation, Guadarrama adopts a longitudinal ventilation system, with ventilation shafts in the portal areas, while Gotthard-Base has a semi-transverse ventilation system, with four very deep and powerful ventilation shafts, two at the ends of each of two intermediate MFSs (Multi-functional Stations).

For fire fighting, the two tunnels also adopt conventional systems, with fire brigades entering the tunnel, through an emergency train (in reverse mode in the Gotthard-Base).

The evacuation of users in an incident situation is carried out, in both tunnels, by means of rescue trains that circulate in the non-incident railway gallery. However, the distances to be covered by a train on fire can be quite large, 28 km in Guadarrama and 23 km in Gotthard-Base (northern section), which will take too long, almost 20 minutes (average speed of 80 km/h, as usually admitted).

4.2.4   Mono-tube tunnels with two single-track galleries

Inexplicably, so far, there are not many cases of mono-tube double gallery single-track railway tunnels built. Below is presented an impressive case, the Groenehart Tunnel, in The Netherlands.

Groenehart Tunnel, The Netherlands

The Groenehart tunnel is an integral part of the HSRL (HSL-Zuid) that extends through Belgium and The Netherlands, located between Rotterdam and Amsterdam, slightly south of Schiphol airport. It has been in operation since its opening in 2005, with trains at commercial speeds of 300 km/h, although it is prepared for train speeds of 350 km/h.

In that area the works had to be underground due to environment restrictions, since the line crosses the protected area of the "Groene Hart" ("green heart"), with access being absolutely limited to site investigations.

The project consists of a 7.2 km long tunnel with two access ramps 740m and 710m long, on the north and south side respectively.

The local geology is homogeneous, with a top layer of 10 to 15 meters thick of soft clay, on layers of sand with high permeability, which potential stability problems (face stability and buoyancy). The tunnel runs 25m below ground level and crosses two major aquifers.

The original project foresaw a tunnel with two parallel TBM tubes, each for a track, with 9.5m inner diameter each, and air sections of 60 m^2 (figure 4.13) (Bouygues 2001), connected together every 300m via cross-passages, to which users would have access to through emergency walkways arranged along the inside of the tubes.

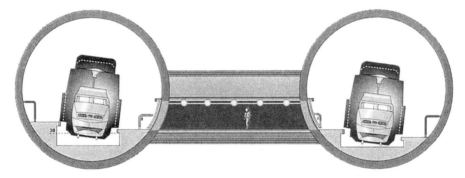

Figure 4.13.   Groenehart Tunnel. The original TBM solution

The contract was of the Design & Build type and expressly called for innovative techniques to limit the risk profile, in particular for the tunnel connections.

An alternative solution was then developed by Bouygues, consisting of a TBM single-tube dual-track tunnel with a separating wall in the middle and eliminating the cross-passages (figure 4.14) (Bouygues 2001), which was eventually the choice of the Dutch Administration.

The inner diameter of the tube is 13.30m, providing 50 m^2 air sections in each track, to allow train speeds of 350 km/h.

Precast segments with 0.60m thickness were adopted (higher than usual, due to the particular bad soil conditions) and injections of 0.185m thickness, which led to an excavation diameter of 14.87m, the world record for TBMs at the time. A Slurry-type TBM, supplied by NFM Technologies, was used (Aristaghes 2003).

The contract was signed in December 1999 and the works took 41 months, from September 2001 to May 2005, of which 32 months for the construction of the tunnel (225 meters of tunnel per month, as average). The contract price was EUR 430 million, which represents a relatively low cost ratio of EUR 59 million per km of tunnel (including the cost of the access ramps).

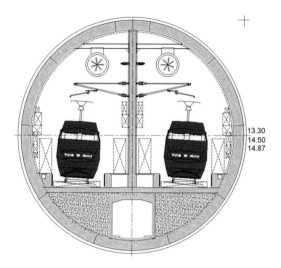

Figure 4.14.   Groenehart Tunnel. The alternative TBM solution

The functionality in operation and safety concepts of the tunnel were fully revisited by the contractor in order to satisfy the Dutch administration. Thus, the number of the evacuation outlets in the separating wall (escape doors) has been doubled compared to the initial project (every 150 meters, instead of 300 meters). In addition, a technical gallery for cables and pipes was built below the tracks, to be accessed through access galleries created in the outer wall (lining) of the tunnel (figure 4.15) (Aristaghes 2003).

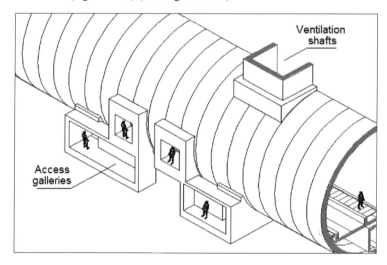

Figure 4.15.   Groenehart Tunnel. Access galleries to the technical gallery and ventilation shafts

The separating wall was provided with openings in the upper part for air pressure release, which allow the reduction of the cross-sectional area of the railway galleries to $50 \ m^2$, keeping the comfort conditions inside the tunnel, which will be quickly closed during a fire.

For ventilation, longitudinal reverse jet fans were disposed along the upper part of the two railway galleries, supplemented by two ventilation shafts. During a fire, jet fans on both sides continue working in order to prevent excessive heating of the concrete on the site.

In the case of fire or train failure, passengers will leave the incident gallery through the escape doors and take the side railway gallery, to be rescued later by another train. However, the escape doors open directly to the side railway gallery, which can make difficult the control of smoke entering the non-incident gallery during a fire.

Fire fighting is through conventional means, by fire brigades that enter the tunnel through the tunnel portals.

According to those involved on its construction, the project was a great success, with the alternative solution having many advantages over the basic solution, namely in terms of environment protection, satisfaction of safety and economy requirements and risk reduction.

As summarized by the Contractor, the alternative was more economical and attractive (Martin, 2009): "One tunnel, one TBM, one precast yard, one construction team that is significant in term of logistics; elimination of the cross-passages, to be built under freezing in difficult conditions and on critical path; reduction in the width of the entry and exit access trench; improved control and maintenance of electrical cables and water pipes, located in the gallery below the tracks".

After 15 years in operation, no relevant event took place in the tunnel so far.

Remarks

As mentioned, the Groeneheart Tunnel, based on the mono-tube with single-track separated galleries concept, is an alternative to the dual-tube tunnels with single-track tubes and the only railway tunnel based on that concept, built so far. However, according to those involved, the project was a great success, namely in terms of environment protection, economy and compliance with safety requirements.

Compared to the basic solution, it allowed, inter alia, the elimination of the cross-passages along the tunnel and a significant simplification of the access ramps. The execution of openings in the upper part of the separating wall also allowed a significant reduction in the cross-sectional area of the railway galleries and, consequently, in the diameter of the tunnel.

The safety in operation concept is also quite simple, however, in the event of a fire in one of the traffic galleries, the escape doors open directly to the adjacent gallery, which makes it difficult to control the entry of smoke in the non-incident gallery.

4.3   TBM ROAD TUNNELS

4.3.1   Current concepts in TBM road tunnels

Road tunnels have long been formed by single tubes, with the two traffic directions placed side by side.

However, in order to satisfy the increasingly demanding safety requirements, in the mid-1990s the **dual-tube tunnel concept** (formerly developed for railway tunnels) started to be used, in which the tunnel consist of two separate TBM tubes, each for a traffic direction, and one or more technical galleries at the bottom, connected by cross-passages (figure 4.16a).

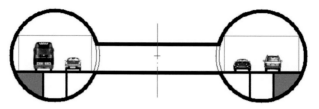

a) Dual-tube unidirectional tunnel

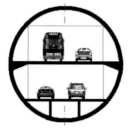

b) Mono-tube double-deck unidirectional tunnel

Figure 4.16.   Current concepts in TBM road tunnels

Subsequently, a new concept was developed and also started to be used, the **mono-tube double-deck tunnel concept**, in which the tunnel is formed by a single tube with two super-imposed traffic galleries, each one for a traffic direction and one or more technical galleries at the bottom (figure 4.16b).

Below are noteworthy examples of road tunnels built on the basis each of these concepts, focusing on aspects such as geometry, construction methods and technical systems, namely ventilation and fire fighting, and safety in operation.

4.3.2   Dual-tube unidirectional tunnels

Many examples of dual-tube unidirectional TBM road tunnels have been built in last two decades. Because of their peculiarities, the following cases are presented: the Westerschelde Tunnel, in the south Netherlands; the M30 South By-pass Tunnel in Madrid, Spain; and the Changjiang Tunnel in Shanghai, China.

Westerschelde Tunnel, The Netherlands
The Dutch province of Zeeuwsch-Vlaandren is isolated from the other parts of the country by the Westerschelde entrance to the Belgian port of Antwerpen.

Projects for a permanent link in the area have been developed since the 1930s. In the 1990's, a decision was made to make this connection through the so-called Westerschelde Tunnel, a 6.6 km underwater road tunnel (Wikipedia 2021).

The Westerschelde Tunnel provides a fast and economic route for around 27 000 vehicles a day between Ellewoutsdijk in Zuid-Beveland on north, and Hoogedijk (near Terneuzen) in the Zeeuws-Vlaandren region on south, close to the Belgium border. However, some restrictions on the traffic were imposed, namely LPG-cars or trucks with heavily toxic materials are not allowed in the tunnel.

KMW group, led by Philipp Holzmann, were awarded the contract for the design, construction and maintenance of the tunnel for the first ten years of operation. The total cost of the tunnel was EUR 726 million (including the costs of the access ramps on both sides) and was opened to traffic in March 2003.

The tunnel consists of two parallel TBM tubes, each with two lanes 3.5m wide and 4.50m clear height, connected by 26 cross-passages, spaced 250 meters (figure 4.17) (Hoonaard 2003). It runs up to 60m below sea level (water pressure of 6 bar).

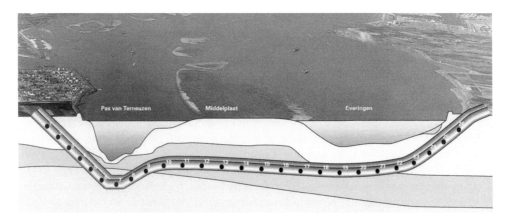

Figure 4.17. Westerschelde Tunnel. Location and longitudinal profile

The inner diameter of each tunnel tube is 10.10m. Precast segments 0.45m thick and 2.0m wide, and an annular gap of 0.165m were adopted, so the excavated diameter of the tubes is 11.33m, the world record at the time (figure 4.18) (Hoonaard 2003, adapted).

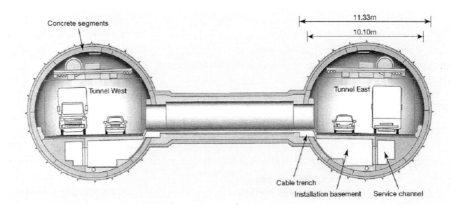

Figure 4.18. Westerschelde Tunnel. Cross-section

Both road galleries are provided with edges 0.70m wide on each side, making a total road width of 8.40m, where the efficiency index of each tube presents an interesting value of 0.47 (see, Chapter 2). However, both tunnel tubes do not have emergency walkways, neither emergency lane.

A service gallery below the road lanes houses all the wiring for the tunnel outfit, such as air control, signalling, supplies, and safety and communication systems. Additional drainage devices are found at the bottom of each tube.

The cross-passages have 12m in length and 2.5m x 2.7m cross-section.

A particular challenge during the construction of the tunnel was to bore through the very sticky highly-plasticized Boomse clay. Two TBMs of the Mixshield type, supplied by Herrenknecht AG, were used. They started from an artificial polder in Hoogedik and bored through to concrete caissons 20m deep at Ellewoutsdijk. The advance rate of the TBMs was 12m per day, on average.

The tunnel tubes were, at the time, the deepest ever driven worldwide, so they posed great challenges in terms of layout, processing and construction of the machines. An open spoke-type cutting wheel with rim, an active central cutter with own flushing circuit, agitators and a rolling crusher in front of the suction pipe enabled optimum material transport.

The two tunnel tubes have a total concrete lining of 53 000 segments. As local roads were not able to cope with the huge amount of material transport, all major items were supplied via ship and handled at an own landing facility.

The production of the precast segments took place in a twin plant on site. All material going into the tunnel was transported exclusively by rail. Eight locomotives and 80 wagons operated out of the logistics centre in Terneuzen and were responsible for the transport of the people and materials needed for the construction of the tunnel. In all, about 275 000 ton of concrete were transported in all. Significant part of the 1.6 million m^3 of spoil produced by the excavation was reused for concrete.

The tunnel was designed on basis of a designed service life, in this case, 100 years. The intended functions and the required reliability were defined and then determined which functions would be adversely affected by concrete deterioration.

The tunnel tubes lining was provided with an internal fire protection coating, designed for a fire load of 1350°C, which was applied by automatic spray robots.

The cross-passages were built using ground freezing methods. A 2 meters thickness frosted "tube" was created around each cross passage by drilling 22 holes for freezing. After excavating the inside of the "tube", an external wall of 0.30m thickness made of shotcrete was built and then an internal reinforced concrete wall 0.40m thick was cast (figure 4.19) (Hoonaard 2003, adapted).

The refrigerating medium used was a calcium chloride solution, applied at a temperature of −37°C. Each "tube" took about 30 to 45 days to freeze, depending on the specific ground conditions in each case.

Both tunnel tubes are equipped with a longitudinal ventilation system by means of jet fans. In case of a fire or when smoke is created, ventilation is activated in the same direction as the traffic.

Fire attack in the tunnel tubes is carried out by conventional means, through fire brigades. In the event of an incident inside one of the tunnel tubes, emergency services (rescue services and fire fighters) access the site through the non-incident tube and the cross-passages,

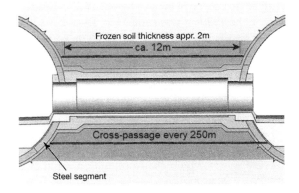

Figure 4.19. Westerschelde Tunnel. Execution of the cross-passages

using special emergency electric vehicles, similar to those used in the Channel Tunnel (see, figure 4.5).

M30 South By-Pass Tunnel, Madrid, Spain

The M30 highway in Madrid surrounds the central districts of the city, being the busiest Spanish road, with an AADT of 75000 vehicles. It has 32.5 km in length (figure 4.20) (Wikipedia 2021c, adapted) and three lanes in each direction, complemented in some parts by auxiliary lanes. It connects to the main Spanish radial national roads that start from Madrid.

Figure 4.20. M30 Ring in Madrid, Spain. Location of the SBP Tunnel

The construction of the ring began in the 1960s and in the 1970s the most important section was opened to traffic; in the 1990s it was completed. From 2005 to 2008, major upgrading works took place, and a significant portion of the southern part started to run underground, one of them so-called South By-Pass (M30 SBP) Tunnel. However some limitations were imposed in the operation: the traffic of dangerous heavy vehicles is prohibited, as well as that of vehicles above 7.5 tons. The speed limit is 70 km/h.

The tunnel has two different sections: one, 2.0 km long, was built using the cut-and-cover method, and the other, 3.6 km long, was built using the TBM technique.

The TBM section is formed by two parallel tubes, connected by cross-passages, every 200 meters. The inner diameter of the tubes is 13.45m, which led to excavated diameters of 15.00m in the South tube and 15.20m in the North tube, a world record at the time, and the use precast segments with 0.70m thickness (figure 4.21) (PIARC 2011a, adapted).

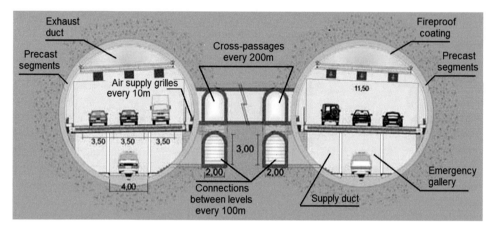

Figure 4.21.   M30 SBP Tunnel. Cross-section

The cross-section of each tube comprises a road platform, housing three lanes 3.50m wide and edges 0.50m wide on both sides, making a total road width of 11.50m, in addition to 0.70m wide emergency walkways on each side. The vertical clearance of the road galleries is 4.50m, so the efficiency index of each tube has a modest value of 0.36.

Since most of the excavation was in soft soil (little hard rock), EPB type TBMs were used. Overburden vary up to 75m above the tunnel axis and water pressures up to 6 bar. The TBM for the North tube, supplied by Herrenknecht AG, had two cutting wheels, allowing them to operate in clockwise and counter-clockwise rotations, independently of each other.

Cross-passages are located every 200m and are connected by stairs to escape routes underneath each roadway. Every 600m the cross-passages are designed to allow vehicles to pass from one tube to the other. There are also stairs for the exclusive use of fire fighters that connect the emergency gallery and the roadway. The cross-passages are closed by escape doors and pressurized, being built using conventional excavation methods.

Ventilation systems of the tunnel tubes are of the transverse type. In the upper part of the cross-section of each tube a ventilation duct to exhaust polluted air was arranged. Underneath the roadway platform there is a gallery with a vertical clearance of 3.6m, which serve as emergency route and fresh air supply duct (figure 4.21).

Four ventilation shafts, spaced 600m on average, exhaust polluted air (and smoke) and provide fresh air, through axial fans. Polluted air is exhausted by grilles placed on the upper duct, while outside air is supplied through grilles placed at each side of the roadway.

In the regular ventilation, the axial supply fans and the axial exhaust fans come into action whenever the sensors located inside the tunnel indicate that the pollution is higher than the allowable limits.

In the fire ventilation strategy, in the section in which a fire is located the exhaust and the supply fans are switched to emergency mode, while in the adjacent sections exhaust fans are stopped the supply fans operate at maximum flow. Once the evacuation of users is finished, the system will operate with longitudinal ventilation using the high volume extraction of the ventilation shafts.

An overpressure of 50 Pa in the cross-passages and in the emergency gallery underneath the roadway prevents smoke from entering in those galleries.

Fire fighting is carried out by fire brigades, who enter the tunnel in fire trucks through the emergency galleries.

Changjiang Tunnel, Xanghai, China
The Changjiang Tunnel, also known as Shanghai Yangtze River Tunnel, in Xanghai, China, is a road tunnel that connects the Pudong district of Xanghai to the Changxing Island, at north, crossing the South Channel of the Yangtze River estuary, as part of the Chongming Crossing, which connects Xanghai to the Chongming Island and continues north, completing the Yangtze River crossing in the area (Figure 4.22) (Huang 2009, adapted).

Figure 4.22. Changjiang Tunnel. Location

The Chongming crossing was planned on the 1990s, being the preliminary design of the project approved by the authorities in 2005. In April 2009, the tunnel was opened to traffic. The cost was the equivalent to EUR 1200 million (EUR 134 million per km).

Figure 4.23 (Huang 2009, adapted) shows the geological conditions along the tunnel axis. The ground is composed essentially of alluvial deposits: alluvial clay, sandy silt, clay silt, etc. Unfavorable geological conditions were experienced, such as liquefied soil, quick sand, piping, shallow gas (methane), confined water, etc. The ground surface is mostly flat, with a normal elevation of 3.50m. The site is classified as high seismic intensity, 7.0.

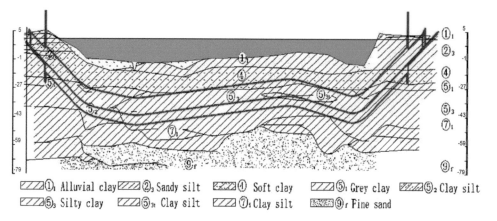

Figure 4.23.   Changjiang Tunnel. Longitudinal geological profile

The tunnel is 8.95 km long, of which 7.5 km are of the TBM type, at a depth of up to 65 meters. The design speed in the tunnel is 80 km/h. The design service life is 100 years.

The TBM tunnel section is similar to M30 SBP Tunnel, consisting of a dual-tube, each with an inner diameter of 13.70m and housing 3 lanes with a total traffic width of 12.75m (figure 4.24) (Huang 2009, adapted). The clear height is 5.20m, so the tunnel tubes have an efficiency index with an interesting value of 0.45.

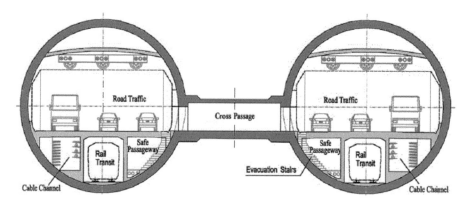

Figure 4.24.   Changjiang Tunnel. Cross section at cross-passages

Each tunnel tube has at the top a large smoke duct and at the bottom galleries for an express train placed in the middle, and channels for cables and pipes, and a pedestrian passageway accessed by stairs, at the sides.

The road galleries of the two tubes are connected by cross-passages about 15m long, 2.1m high and 1.80m wide, arranged every 830 meters. Between each two cross-passages there are three evacuation stairs connecting the upper and the lower levels of each tube, to access the passageway.

The lining of the tubes is formed of precast segments 0.65m thick and 2.0m wide, each ring being constituted by 10 segments, including the "key".

The tunnel tubes were excavated by two Mixshield TBMs, supplied by Herrenknecht AG and assembled in place by its local associate, STEC (Shanghai Tunnel Engineering Co.). The excavated diameter is 14.53m, the largest TBMs in the world at the time (Kühn 2006).

To allow the change of tools at the cutting wheel in safe conditions, the six cutting wheel arms of the machines were accessible from within the shield under atmospheric pressure. The excavation of each 7.5 km tube took about 24 months.

The cross-passages were built as in the Westerschede Tunnel, by locally freezing the soil and making excavations by the mining method. The freezing holes were made on both sides.

The ventilation system in each tunnel tube is of the semi-transversal type, combining longitudinal jet fans with transversal ventilation induced by large ventilators arranged in ventilation shafts located at the portals and using the upper smoke duct (Shen 2008).

During normal operation and congested condition, jet fans are turned on to discharge the polluted air to out of the tunnel tube, in the direction of the road traffic. They are 1.0m in diameter and are suspended from the smoke duct, being arranged in groups of three, spaced 280 meters apart.

When a fire accident occurs inside one of the road galleries, the ventilator in the shaft is turned on to ventilate the smoke to the side of the fire source, while users evacuate towards the fresh air.

In these situations, users escape to the other road gallery through the cross-passages, where they will be later rescued. If there is no cross-passage around, users will descend the stairs to the passageway on the lower level, through openings on the inner side of the floor (see, figure 4.24), which are covered by steel plates. Then they will go up to take the nearest cross-passage. In no case it will be allowed vehicles to park on those plates.

The fire fighting system is quite complex, since fire can occur on the road galleries, on the upper floor, and also on the train gallery, on the lower level.

On the road galleries, fire fighting relies on three conventional levels: firstly, by the affected users; secondly, by the tunnel operators; and thirdly, by the fire brigade, which will accede to the site through the passageway and the stairs.

In the event of fire in the train gallery, passengers will be evacuated to the upper floor, using the passageway and the stairs, while emergency personnel go down to the lower level, to deal with the situation.

Remarks

The three dual-tube road tunnels presented, Westerschelde, M30 SBP and Changjiang, are impressive tunnels, the first with two lanes in each direction and the last two with 3 lanes in each direction. The lay-outs are similar, although presenting significantly different efficiency indexes, from 0.36 on the M30 SBP Tunnel to 0.45 on the Chanjiang Tunnel and 0.47 (the highest value) on the Westershelde Tunnel, with an average value of 0.42.

Ventilation and fire fighting systems have similarities, although with some significant differences. Westerschelde is equipped with longitudinal ventilation by jet fans, M30 SBP uses a transverse ventilation system, with ventilation shafts arranged along the tunnel, and Changjiang uses a semi-transverse system, with longitudinal jet fans complemented by a longitudinal duct and ventilation shafts in the portal areas. Fire fighting is identical on the three tunnels, being carried out by conventional means, through fire brigades.

The safety concepts (safety in operation) are also similar, being, in incident situations, the users rescued through the non-incident road gallery, to which they have access through the systems of cross-passages and stairs (in the two last cases).

4.3.3   Mono-tube double-deck unidirectional tunnels

There are more and more mono-tube double-deck TBM road tunnels built, although some of them with limitations on the height of the vehicles allowed. The following noteworthy cases are presented: the A86 Duplex Tunnel, in Paris, France; the Eurasia Tunnel, in Istanbul, Turkey; and the SR99 Tunnel, in Seattle, USA, as well the Orlovski Tunnel, planned for Saint Petersburg, Russia, although it was not built.

A86 Duplex Tunnel, Paris, France
The A86 Duplex Tunnel, generally simply called Duplex Tunnel, is located west of Paris on the A86 highway, between Rueil-Malmaison and Versailles Pont-Colbert (Velizy), filling the last link of this route, the second circular ring around Paris (figure 4.25) (PIARC 2011b (adapted). The speeds in the tunnel are limited at 70 km/h (Cofiroute 2009).

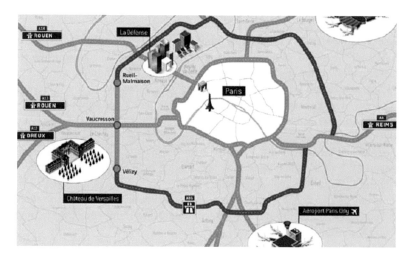

Figure 4.25.   Duplex Tunnel. Location

The tunnel is a mono-tube TBM tunnel about 10 km long (one of the longest road tunnels in the world) housing two superimposed decks with unidirectional traffic. It is reserved to light vehicles up to 2.0 meters in height and weight up to 3.5 tons. Vehicles powered by LPG or LNG, with engine smaller than 50 cm^3 or without a licence plate are prohibited in the tunnel, as well as dangerous goods vehicles (PIARC 2011b).

The tunnel was built on a concession basis by Cofiroute, a subsidiary of Vinci Autoroutes, and tolled until the end of the concession period, which ends in 2086. It was completed in 2011 at a cost of EUR 300 million (EUR 30 million per km).

Tunnel alignment is quite complex. In addition to geological constraints (limestone and marl in the end sections and sand in the intermediate section), it had to take into account the presence of interchange nodes, particularly with the A13 highway. It starts near the Seine River in Rueil-Malmaison and goes up a slope until reaching the A13 highway. Then go down a slope at Viroflay and up another slope at Pont-Colbert in Versailles (figure 4.26) (PIARC 2011b (adapted). The tunnel overburden varies between 15 and 90 meters.

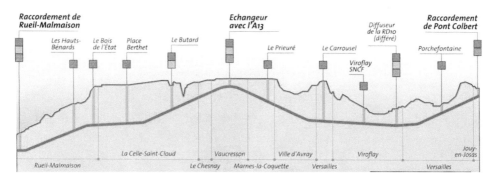

Figure 4.26. Duplex Tunnel. Longitudinal profile

As mentioned, the tunnel comprises two decks, each with unidirectional traffic: the external traffic of the A86 (southbound) is located on the lower floor and the internal traffic of the A86 (northbound) is located on the upper floor (figure 4.27) (PIARC 2011b (adapted).

Each deck has two 3.0m wide traffic lanes, an emergency lane 2.5m wide and an inner edge 0.22m wide, making a total road width of 8.72m. The emergency lane is on the right side of the slow lane on the lower floor and on the left side of the fast lane on the upper floor.

The inner diameter of the tunnel is 10.40m and the height of the road galleries is 2.57m with a free height of 2.0m, so the efficiency index of the tunnel is 0.41.

Precast segments 0.40m thick were used. Due to the so variable soil conditions, the tunnel was built by a TBM functioning in two modes, EPB and Mixshield, with 11.57m excavated diameter, the largest TBM so far designed, when the tunnel was planned, 15 years before construction started.

Recesses (vertical access galleries), pressurized and smoke proof with fire resistance of two hours are disposed every 200m (figure 4.28) (PIARC 2011b (adapted). These recesses, equipped with stairs, connect the two road galleries, providing emergency exits in an incident situation in one of the traffic galleries.

The tunnel is also provided with 13 complex emergency access shafts located along the tunnel, equipped with elevators, the distance varying between 500m and 1200m (figure 4.29) (PIARC 2011b (adapted). These shafts provide emergency personnel with access from the surface to the traffic galleries, enable them to help users to be rescued out of the tunnel.

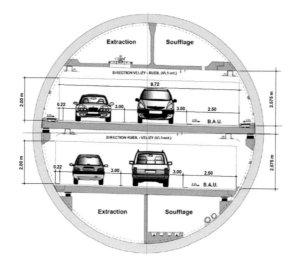

Figure 4.27.   Duplex Tunnel. Cross-section

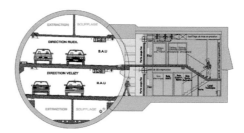

Figure 4.28.   Duplex Tunnel. Safety recesses

The safety recesses and shafts were built by locally cutting the precast segments of the lining and building reinforced concrete walls insitu, after the excavations were carried out.

Tunnel ventilation is a combination of transversal and longitudinal systems, consisting of six ventilation shafts (to inject fresh air and extract polluted air) connected to fresh air injection ducts and extraction ducts, located below and above the traffic galleries (see, figure 4.27), through injection and extraction dampers (8m spacing). Longitudinal jet fans are also arranged along the tunnel.

The purpose of the ventilation system is, in the event of a fire, to prevent smoke from spreading into the traffic space. Therefore, the air flow is maintained in the direction of traffic to prevent smoke from layering upstream of the fireplace.

The smoke is extracted by the damper closest to the fire. If the fire spreads beyond this damper, an air flow in reverse is generated. The ventilation systems for the escape routes are independent from those of the traffic galleries, being pressurized.

For fire fighting, an FFFS (Fixed Fire Fighting System) with sprinklers was installed throughout the tunnel. In the event of a fire, this system sprays water in the form of fine droplets for 30 minutes, over an area of about 100m in length, centered on the fire. Its main

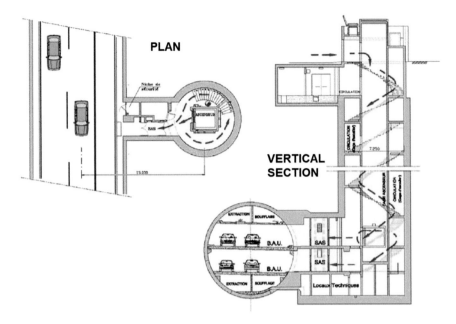

Figure 4.29. Duplex Tunnel. Emergency access shafts

objectives are: (i) to limit the development of the fire, (ii) to facilitate the intervention of the emergency services, (iii) to improve the survival conditions downstream of the fire.

In the event of a fire, both entrances of the tunnel will be closed automatically. The rescue services and the fire brigade then enter the tunnel using special low-height vehicles (two fire brigade vehicles and one ambulance permanently available at the ends of the tunnel).

Despite a great controversy over its safety, in 10 years of service, no major incident, with or without fire, has occurred in the tunnel so far.

Eurasia Tunnel, Istanbul, Turkey

The Eurasia Tunnel (officially Bosphorus Strait Crossing) is part of the Eurasia Tunnel Project, a road link that connects the Asian and European sides of Istanbul, Turquey, in a very sensitive archaeological area and going very deep under the seabed, with an aim to relieve Istanbul's transcontinental traffic pressure. It started operation in December 2016.

The tunnel route was chosen after extensive feasibility studies carried out under the responsibility of the Turquish government that announced an international tender with BOT (Build-Operate-Transfer) model. Tender was won by the ATAŞ consortium, formed by Yapi Merkezi from Turkey and SK E&C from South Korea in February 2011, with the "Implementation Contract" taking effect in January 2013.

The Eurasia Tunnel Project is 14.6 km long, serving along the Kazlıçeşme-Göztepe route, at the Southern end of the Istanbul Strait. It includes a 3.4 km TBM mono-tube double-deck tunnel, 2 km of NATM tunnels, and 1 km of cut-and-cover tunnels, in addition to the road

arrangements along the access roads on both sides (figure 4.30) (DLH 2012). The speed limit
in the tunnels is 70 km/h. The TBM tunnel goes 106 meters beneath the seabed.

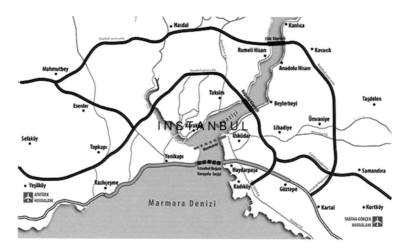

Figure 4.30.   Eurasia Tunnel. Location

The project was completed in 47 months, almost 8 months prior to the construction time
defined in the contract, with a global investment of about EUR 1100 million. It will be
transferred to the public domain after 25 years of operation (WSP 2021). In 2015 it was
awarded the "Tunnelling Project of the Year" by ITA (International Tunnelling Association).

The TBM tunnel has two superimposed road galleries, with two lanes 3.0m wide each and
a 1.2m wide emergency lane, making total road widths of 7.2m each, designed for passenger
vehicles and small vans with a vertical clearance of 2.8 meters. At the base there are several
galleries for services and ducts for cables and piping. The inner diameter of the tunnel is
12.0m, thus the tunnel presents a modest efficiency index of 0.36 (figure 4.31) (DLH 2012).

Precast concrete segments 0.60m thick and 70 MPa strength, and a 0.25m gap for injections
were used, thus leading to an excavated diameter of 13.70m.

Each tunnel ring is formed of 9 segments 2.0m wide, including a key, double reinforced,
right and left tapper, double gaskets, bolted in both directions, shear keys and sliding rods.

The floor slabs are supported on concrete corbels cast in-situ from reinforcement bars
protruding from the inner face of the precast segments.

The construction of the tunnel had to face two major challenges: the need to go very deep
(10.6 bars of water pressure) and the complex geological conditions of the alignment, located
in an active seismic area, about 17 km north of the Anatolian Fault. It was excavated with a
TBM of the Mixshield type, supplied by Herrenknecht AG, the "first" worldwide prepared
to operate at pressure of 12 bars.

Most of the length of the tunnel was excavated in rock formations, also crossing softer
formations on the Asian side, in an extension of about 500m, composed mainly of sandy clay.
It has 5% gradients on the ramps on both sides (figure 4.32) (Davis 2017, adapted).

In order to reduce the stresses in the tunnel lining during a seismic event, two sophisticated
flexible seismic joints (rings), with displacement limits of ±50 mm for shear and ±75 mm

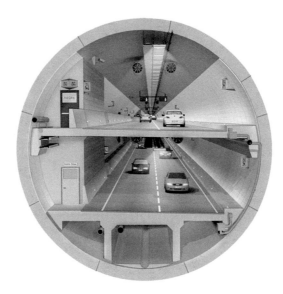

Figure 4.31. Eurasia Tunnel. Configuration

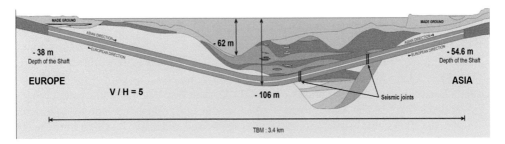

Figure 4.32. Eurasia Tunnel. Geological profile

for extraction/contraction, were adopted in the transitions between the rock formations and the softer soil stretch. These joints were designed for the specific ground conditions and implemented for the first time in TBM tunnels under such high pressure (figure 4.33) (Schmid 2020).

The 3.4 km long TBM tunnel was completed in less than 16 months. During the excavation, hyperbaric maintenance and repair operations were carried out for four times, operations that were successfully performed by specially trained divers and the TBM crew. One occurred precisely in the lowest depth zone, at 10.6 bar pressure which was also a "first" in TBM tunnelling.

Ventilation is carried out through longitudinal ventilation via jet fans disposed in the ceiling of each road gallery, complemented by ventilators arranged in two ventilation buildings disposed at both tunnel portals. The fire fighting is carried out by fire brigades, who enter the tunnel in fire trucks through the tunnel portals.

Figure 4.33.   Eurasia Tunnel. Seismic joints

The tunnel is provided with escape doors every 200 meters, which give access to pressurized vertical access galleries equipped with stairs, placed on one side inside the tunnel, which provide shelter and escape route for users to the other level of the tunnel, to be later rescued in an incident situation (see, figure 4.31).

SR99 Tunnel, Seattle, USA

The SR99 (State Route 99) Tunnel, in Seattle, USA, was built as part of the Washington State Department of Transportation's (WSDOT) Alaskan Way Viaduct Replacement Program (Wikipedia, 2021d). The Alaskan Way Viaduct was an aging double-deck highway structure, built in the 1950's, that had deteriorated due to age as well as damage resulting from the 2001 Nisqually earthquake.

In December 2010, WSDOT awarded the SR99 TBM Tunnel Design-Build Project to STP, a joint-venture between Dragados USA and Tutor Perini Corp (Wallis 2011).

The tunnel was opened to traffic in February 2019. It was designed for all types of traffic, including heavy goods vehicles. Its construction also opened up areas that needed improvement and offers Seattle the opportunity to improve its waterfront.

The tunnel consists of three stretches: the South Approach, the TBM Tunnel and the North Approach. The South and North Approaches include cut-and-cover tunnels and U-sections.

The TBM tunnel is 2.8 km long. It begins south of downtown Seattle in close proximity to the seawall of Elliott Bay, tends north along the existing Alaskan Way Viaduct, then crosses under the Viaduct, traverses under downtown Seattle, and emerges north of downtown Seattle just east of the Space Needle (figure 4.34) (Jiang 2019 (adapted).

The tunnel is a mono-tube double-deck 15.70m inner diameter, with two superimposed road galleries with two lanes 3.33m wide each, a 0.60m wide edge on the east and a 2.40m emergency lane on the west, making a total road width of 9.66m (figure 4.35) (Jiang 2019 (adapted). The vertical clearance is 4.67m, so the efficiency index of the tunnel has the significant value of 0.47.

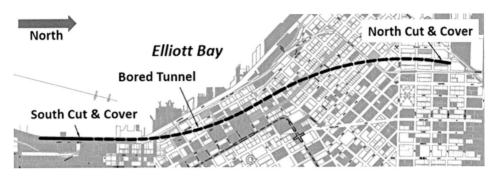

Figure 4.34. SR 99 Tunnel. Location

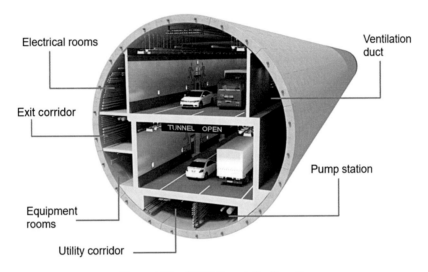

Figure 4.35. SR 99 Tunnel. Configuration

As regard geology, along its alignment, the tunnel traverses through variable glacially over-consolidated soil deposits with high groundwater pressures of up to 5.2 bars (figure 4.36) (Jiang 2019 (adapted). At its lowest point, the tunnel crown is at elevation −29, and it is 65m deep at its greatest depth below grade.

Along the tunnel alignment there are level and elevated roads, railways, sewage tunnels and public and private services, and buildings ranging from single-story to high-rise structures.

The tunnel was excavated by a TBM of EPB type, named "Bertha", supplied by Hitachi Zosen Corporation, with a 17.50m diameter cutterhead, the largest TBM in the world at the time. The rings of the tunnel lining consist of 10 precast segments 0.60m thick and 2.0m wide.

Due to the particular seismic conditions of the site, beyond static analysis, great attention was paid to the verification of the seismic behaviour of the structure (Jiang 2019). Design sections were selected to assess the geologic variability along the tunnel alignment, as well as topographic/geometric variability and building locations.

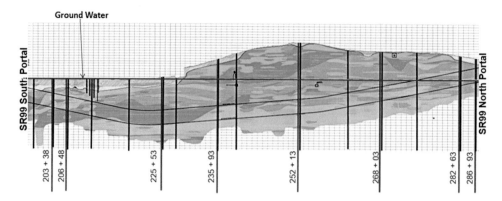

Figure 4.36.   SR99 Tunnel. Geotechnical profile and locations of design sections

As shown in figure 4.36, 15 geologic sections (in blue) were selected for static design and 8 sections (in red) for seismic design. In view of the future development of the area, a significant surcharge was adopted.

Two levels of seismic design were performed. Under the "Expected Earthquake", lining was designed to respond in an elastic manner, in order to prevent water tightness. Concrete compression strain was limited to 0.003, and tensile strain in reinforcing steel limited to 0.020. Under the "Rare Earthquake", the objective was to avoid the collapse of the lining, therefore, inelastic deformations were admitted. Concrete strain was allowed to exceed 0.003, but limited to 0.005, provided that the deformation was predominantly due to flexure.

Analysis included loading from seismic deformations and ground accelerations considering three primary modes of deformation: ovalization, axial deformations and curvature deformations. The lining segment gaskets were also evaluated for water tightness under this maximum ovalization, as well as the maximum longitudinal curvature.

As regard construction, there was a significant setback to face. The TBM began drilling in July 2013 and only five months later it halted drilling due to overheating of some of its components (Johnson 2014). After two years of repair, the TBM resumed tunnelling in February 2016 and finally broke through into the receiving shaft in April 2017.

The construction of the interior structure of the tunnel also put several challenges. It is comprised of two continuous corbels supporting a series of 200m long frame systems of walls and slabs, detailed to expand and contract longitudinally (figure 4.37) (Jiang 2019 (adapted).

The lower roadway walls are primarily pin connected to the corbels below and fixed to the upper roadway slab. The upper walls are pin connected on each end, detailed to accommodate transverse seismic deformation of the frame and the tunnel lining.

The slabs of the electrical rooms and of the exit corridors are cantilevered from the interior walls; the cantilevered slabs and upper roadway walls are clear enough from the tunnel lining to accommodate the predicted ring ovalization due to the seismic ground movement.

The interior structure was built in a factory-line style system (figure 4.38) (Jiang 2019 (adapted). At the tail of the TBM, dowels were drilled into the inner side of the precast segments in blocked locators prior to casting. Behind the tail, a rail just inside each face was installed, to support the traveling form system gantry.

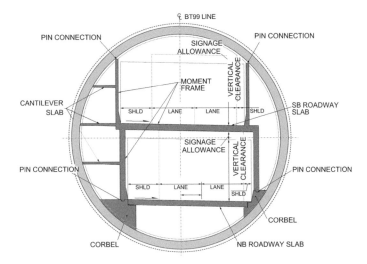

Figure 4.37.   SR99 Tunnel. Cross-section

a) Corbel construction   b) Skip forming for upper slab

Figure 4.38.   SR99 Tunnel. Construction of the interior structure

End-forms for each corbel in 15m intervals were built, the lining surfaces were prepped, corbel reinforcement cages were trucked in and placed, embedded conduits installed, and traveling forms were lowered into place and concrete pumped in by ready-mix trucks. All the while, the invert of the tunnel was kept free of obstructions for passage of the ring segment hauler and shift changes of workers.

Construction of the walls was similar, with skip forming, pre-assembled reinforcing cages, and cast-in-place with a rail supported traveling formwork.

The ventilation system of the tunnel is based on longitudinal jet fans for normal operation, supplemented by an exhaust system to operate in emergencies, constituted by four giant fans connected to chimneys located on both tunnel portals and a ventilation duct disposed along one side of the tunnel's traffic galleries (see, figure 4.35).

In the event of a fire inside the tunnel, the giant exhaust fans act like vacuums, pulling smoke through vents arranged in the walls of the traffic galleries. The smoke travels the length of the tunnel ventilation duct, up and out through the chimneys.

For firefighting, an FFFS (Fixed Fire Fighting System) based on water mist sprinklers was installed in the road galleries of the tunnel, divided in 100 sections 28m long each, with only those in the fire area being activated.

For the evacuation of users in an incident situation, the tunnel disposes of 15 emergency refuge areas located every 200m and exit corridors leading to the portals.

Orlovski Tunnel (Project), Saint Petersburg, Russia

The Orlovski Tunnel Project is a mega road tunnel planned to be built in Saint Petersburg, Russia, to overcome a significant local constraint. The Neva River crosses Saint Petersburg from the Baltic Sea to the harbour area, east of the city and the Ladoga Lake. Thus, all balance bridges are opened at certain hours, mostly during night time, to allow big ships coming from the Baltic Sea to reach the harbour. During these periods, the Neva River can't be crossed.

The Orlovski Tunnel Project would improve this traffic situation and connect the separate areas of the city 24 hours a day. The bridges could be kept open for longer periods, so improving ship traffic (figure 4.39) (Bobylev 2012). However, the tunnel was not built.

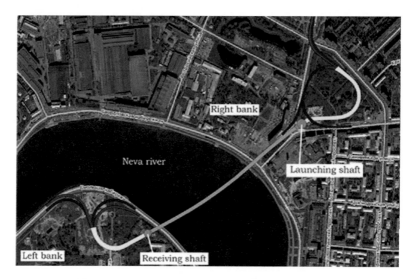

Figure 4.39.   Orlovski Tunnel, in St. Petersburg. Location

The tunnel was to be built as PPP (Public-Private-Partnership), to be refinanced by tolls paid by users. In September 2007 the Concession Agreement was concluded, in June 2010 the contract with the concessionaire NCC, led by Vinci Group, was signed, but in November 2012 it was cancelled by the Russian authorities.

The geology in the area of the crossing consists of quaternary soils, deposed by rivers during melting of the Nordic glaciers at the end of the last ice age. Caused by several advances of the glaciers, the underground is partly over consolidated. On figure 4.40 the longitudinal geological section along the tunnel is presented (Wallis 2012).

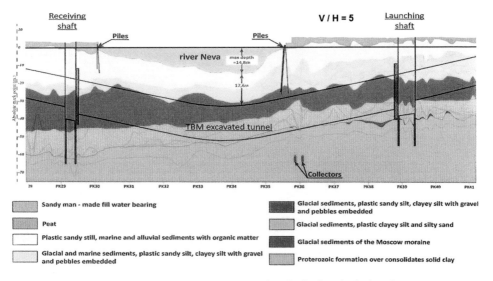

Figure 4.40.   Orlovski Tunnel Project. Longitudinal geological section

The project has to face unique conditions and challenges in the construction of underground structures in urban areas. In addition to an unusual cross-section, it has interferences with the urban environment, the underpass of the river embankments and the very poor soils in the mined section, with low overburden.

The length of the TBM tunnel is 1.0 km, with deep and extensive cut-and-cover approaches, 3 km long on one side and 3.8 km long on the other. It reaches a depth of 50 meters, while the launch and receiving shafts are about 40 meters deep. Due to the characteristics of the ground, it was planned to use a Mixshield TBM, ordered from Herrenknecht AG (Herrenknecht 2011).

The tunnel would allow the circulation of public buses and emergency vehicles, motorcycles, cars and light goods vehicles up to 3.5 tons. The height of the traffic space within the traffic zone would be 4.50m (figure 4.41) (Gruebl 2012).

The inner diameter of the tunnel is 17.25 meters, with two superimposed road galleries housing three lanes each, making a total road width of 11.00m each, so the tunnel efficiency index is 0.42.

Precast segments with 0.70m thickness and 0.30m thick injections were foreseen, so the diameter of the TBM would be 19.25 meters, by far the largest in the world (Gruebl 2012; Wallis 2012). Each tunnel ring is formed by 13 segments 2.20m wide, including the "key".

Inside the tunnel two slabs are built, as well as some partition walls, to accommodate access galleries connecting the two traffic galleries. A service gallery as well as secondary galleries were provided at the bottom of the tunnel, to accommodate piping and cable systems.

Information on the technical systems of the tunnel is not known.

Figure 4.41. Orlovski Tunnel Project. Perspective of the conceptual design

Remarks

The four mono-tube double deck road tunnels presented: Duplex, Eurasia, SR99 and Orlovski (project), are impressive structures, each with two superimposed road galleries, the first three with two lanes in each direction and the last with 3 lanes in each direction (but it was not built). The clear height of the road galleries differs between them; SR99 and Orlovski for vehicles without particular height limitations, and Duplex and Eurasia for vehicles up to only 2.0m and 2.8m high, respectively. Efficiency indexes are also significantly different, from 0.36 in the Eurasia Tunnel, 0.41 in the Duplex Tunnel, 0.42 in the Orlovski Tunnel and 0.47 (the highest value) in the SR99 Tunnel, with an average value of 0.42, by coincidence equal to the average value of the efficiency indexes of the dual-tube road tunnels presented above.

Ventilation and fire fighting systems have similarities, but also significant differences. With regard to ventilation, the three tunnels built are equipped with mixed systems, combining longitudinal ventilation by jet fans with transversal systems with longitudinal ducts and ventilation shafts arranged along the tunnel (Duplex case), or located at the portals (Eurasia and SR99 cases). As for fire fighting, in the Eurasia it is carried out by fire brigades, who enter the tunnel in fire trucks, while Duplex and SR99 are equipped with FFFSs (Fixed Fire Fighting Systems) with water mist sprinklers, arranged along the entire length of the tunnel and divided into sections.

Regarding safety in operation (safety concept) there are significant differences. Duplex and Eurasia are provided with vertical access galleries to allow users to be rescued through the non-incident gallery. In Duplex there are also emergency shafts arranged along the tunnel for the access of emergency personnel to assist in the escape of users. The SR99 is equipped with escape doors that give access to exit corridors arranged along the tunnel, which lead to the tunnel portals.

In Eurasia and SR99, specific provisions were also adopted for earthquake events. In Eurasia, sophisticated flexible seismic joints were adopted in the transitions between the rock formations and the softer ground. However, it will also be necessary to ensure that the tunnel is strong enough to withstand the lateral movements imposed by the ground along that section of the tunnel.

4.4 CONCLUSIONS

Noteworthy examples, some of the largest railway and road TBM tunnels built around the world in the recent times, have been presented above, in which the most advanced tunnel concepts have been adopted.

In relation to railway tunnels, tunnels based on three different concepts were presented: three-tube tunnels, dual-tube tunnels and mono-tube double-gallery tunnels.

As for the **three-tube tunnels**, the Channel Tunnel, between UK and France, is the only tunnel built so far on the basis of this concept. With more than 25 years in operation it fulfilled its mission, but revealing some gaps. It is worth noting the very high cost, about EUR 257 million per km, and the weaknesses in some technical systems, particularly in the firefighting, which led to the installation of a new one, a WFS, however requiring expensive adaptation works. Also the pressure relief system of the railway galleries as well the rescue system of users in incident situations are not completely satisfactory.

As for the **dual-tube tunnels**, the Guadarrama in Spain and the Gotthard-Base in Switzerland, the first for high-speed passengers traffic, and the second for mixed traffic (freight and passengers). Although the global lay-outs are similar, they differ significantly. With regard to cost, although the ground conditions are not very different, in the Guadarrama, it was only EUR 43 million per km, while that of the Gotthard-Base it was much higher, about EUR 149 million per km (more than three times higher), in part due to the very complex technical systems adopted.

With regard to ventilation, the Guadarrama adopts a longitudinal system, with ventilation shafts at the portals, while the Gotthard-Base adopts a semi-transverse system, with very deep ventilation shafts at the ends of two intermediate MFSs ("multi-functional stations"). For fire fighting, the two tunnels adopt conventional systems, with fire brigades entering the tunnel through an emergency train (via reverse mode in the Gotthard-Base).

The evacuation of users in an incident situation is, in both tunnels, by means of rescue trains that circulate in the non-incident railway gallery. However, the distances to be covered by a burning train can be quite large, 23 km in the Gotthard-Base (northern section) and 28 km in the Guadarrama, which will take too long, almost 20 minutes.

As for the **mono-tube dual-gallery tunnels**, the only case known, the Groeneheart Tunnel in The Netherlands, which was built as an alternative to a dual-tube tunnel solution, was a great success. Compared to the basic solution, it allowed, inter alia, the elimination of the cross-passages and a significant simplification of the access ramps. The execution of openings in the upper part of the separating wall also allowed a significant reduction in the cross-sectional area of the railway galleries and, consequently, in the diameter of the tunnel.

The safety concept is also quite simple, however, in the event of a fire, the escape doors open directly to the adjacent gallery, which makes it difficult to control the entry of smoke in the non-incident gallery.

In relation to road tunnels, tunnels based on two different concepts were presented: dual-tube tunnels and mono-tube double deck tunnels.

As for the **dual-tube tunnels**, three cases were presented, the Westerschelde Tunnel in the Netherlands, the M30 SBP Tunnel in Spain and the Changjiang Tunnel in China, the first with two lanes and the last two with 3 lanes in each direction. The general lay-outs are similar, although presenting significantly different efficiency indexes, from 0.36 on the M30 SBP, 0.45 on the Chanjiang and 0.47 (the highest value) on the Westershelde, with an average value of 0.42.

Ventilation and firefighting systems have similarities, although with significant differences. The Westerschelde is equipped with longitudinal ventilation by jet fans, the M30 SBP uses a transverse system, with ventilation shafts arranged along the tunnel, and the Changjiang uses a semi-transverse system, with longitudinal ducts and ventilation shafts in the portal areas, complemented by longitudinal jet fans. Firefighting is identical on the three tunnels, being carried out by conventional means, through fire brigades.

The safety concepts are also similar, being, in incident situations, the users rescued through the non-incident road gallery, to which they will have access through the systems of cross-passages and also of stairs (in the two last cases).

As for the **mono-tube dual-deck tunnels**, four cases were presented: the Duplex Tunnel in France, the Eurasia Tunnel in Turkey, the SR99 Tunnel in USA and the Orlovski Tunnel (project) in Russia, the first three with two lanes in each direction and the last with 3 lanes in each direction (however, not built). The clear height of the road galleries differs between them; the SR99 and Orlovski for vehicles without particular height limitations, and the Duplex and Eurasia for vehicles up to 2.0m and 2.8m high, respectively. Efficiency indexes are also significantly different, from 0.36 in the Eurasia, 0.41 in the Duplex, 0.42 in the Orlovski and 0.47 (the highest value) in the SR99, with an average value of 0.42, the same obtained for the dual-tube tunnels.

As for ventilation and firefighting systems there are similarities, but also significant differences. The three tunnels built are equiped with semi-transverse ventilation systems, combining longitudinal jet fans with longitudinal ducts and ventilation shafts arranged along the tunnel (the Duplex case), or at the portals (the Eurasia and SR99 cases). As for firefighting, in the Eurasia it is carried out by fire brigades on fire trucks, while the Duplex and SR99 are equipped with FFFSs, arranged along the entire length of the tunnel.

Regarding safety in operation there are also significant differences. The Duplex and Eurasia are provided with vertical access galleries to allow users to be rescued through the non-incident gallery. The Duplex is still provided with emergency shafts for the access of emergency personnel to assist in the escape. The SR99 is equipped with escape doors that give access to exit corridors arranged along the tunnel, which lead to the tunnel portals.

In the Eurasia and SR99, specific provisions were adopted for earthquake events. In the Eurasia, sophisticated flexible seismic joints were adopted in the transitions between the rock formations and the softer ground, however, it would be necessary to ensure that the tunnel is strong enough to withstand the lateral movements imposed by the ground along that section of the tunnel.

REFERENCES

ADIF (2021a), *Madrid – Valladolid Line. Guadarrama Tunnel*. Adifaltavelocidad.es, 2021

ADIF (2021b), *Perfil Geologico Longitudinal. Tuneles Alta Velocidad de Guadarrama (Longitudinal Geological Profile. High Speed Tunnels of Guadarrama)*. Adifaltavelocidad.es (in Spanish), 2021

ADIF (2021c), *Tunel de Guadarrama (Guadarrama Tunnel)*. Adifaltavelocidad.es (in Spanish), 2021

Anagnostou G, Ehrbar H (2013), *Tunneling Switzerland*. Swiss Tunneling Society (STS), Zurich, Switzerland, 2013

Aristaghes P et al. (2003), *Groene Hart Tunnel – From Conceptual to Detailed Design*. IABSE Symposium Structures for High-Speed Railway Transportation, Antwerp, Belgium, September 2003

Bobylev N (2012), *Russia Considers Sustainable Urbanization*. TunnelTalk, July 2012

Bouygues (2001), *Avec le Plus Grand Tunnelier, Bouygues Construction, Va Realiser l' Ouvrage de Groenehart, Pays-Bas (With the Largest TBM, Bouygues Construction to Build the Groenehart, Netherlands)*. Communiqué (in French), Bouyges Construction, January 2001

Cofiroute (2009), *Duplex A86 Guide Rueil-Malmaison-A13*. Cofiroute, September 2009

Davis A (2017), *Turkeys Eurasia Tunnel a Major Feat of Engineering Brings Europe and Asia Together*. Highways Today, January 17, 2017

Dennis G (2019), *Channel Tunnel: The 20th Century's Most Expensive Construction Project-but Worth Every Penny*. Rail Magazine, Issue 878, May 8, 2019

DLH (2012), *The Istambul Strait Road Crossing Project*. Project Presentation, 2012

Ehrbar H et al. (2016), *Long Railway Tunnels System Choice – Review for the Future*. GeoResources Journal, N.2, 2016

Fabbri D (2019), *Risk, Contract Management, and Financing of the Gotthard Base Tunnel in Switzerland*. Elsevier, Engineering N. 5, 2019

Geluk J (2006), *High Speed in the Low Lands*. Infraspeed, Netherlands, October 2006

Getlink (2021), *The Channel Tunnel – A Unique Infrastructure in the World*. getlinkgroup.com, 2021

Gruebl F (2012), *Segmental Ring Design. New Challenges with High Tunnel Diameters*. ITA-AITES Muir Wood Lecture 2012, April 2012

Herrenknecht (2011), *Russia Confirms Order for Largest TBM Ever*. Herrenknecht News Release, TunnelTalk, August 2011

Herrenknecht (2016), *Gotthard Base Tunnel*. Herrenknecht AG, June 1, 2016

Holt D (2017), *Channel Tunnel Visit*. wcoff.org, 2017

Hoonaard J, Linde W (2003), *The Westerschelde Tunnel, The Netherlands*. SEI (Structural Engineering International), Volume 13, N⁰ 1, February 2003

Huang R (2008), *Shanghai Yangtze River Tunnel. Theory, Design and Construction*. Taylor and Francis, 2008

Huang R (2009), *Overview of Shanghai Yangtze River Tunnel Project*. Geotechnical Aspects of Underground Construction in Soft Ground, Taylor and Francis, 2009

Jiang Y, Herten A (2019), *SR99 Tunnel in Seattle*. Structure Magazine, January 2019

Johnson K (2014), *Reanimating Bertha, a Mechanical Behemoth Slumbering under Seattle.* NYT (New York Times), August 1, 2014

Kühn A (2006), *Ready to Cross the Yangtze – World's Largest Tunnelling Machine Assembled and Ready in 4 Months in Shanghai.* Unitracc.com, May 29, 2006

Kuhn A (2016), *Innovative Ventilation Solutions for Gotthard.* TunnelTalk, August 25, 2016

Luetsche J (2017), *Safety Concept Gotthard-Base Tunnel.* ERA Training, Budapest, June 2017

Martin O et al. (2009), *Integration of Functionality, Technology and Construction for HSL Tunnels.* Workshop Tunnels for High Speed Railways, Porto, Portugal, October 2009

PIARC (2011a), *Madrid By-Pass Tunnel.* Road Tunnels Manual, PIARC WG5 Complex Underground Road Networks, Case Studies: Appendix 2.16, September 2011

PIARC (2011b), *The Duplex Tunnel A86 in Western Paris Area.* Road Tunnels Manual, PIARC WG5 Complex Underground Road Networks, Case Studies: Appendix 2.8, September 2011

River Tunnel – Theory, Design and Construction, Taylor and Francis, 2008

Reynolds P (2011), *Fire Fighting System Unveiled by Eurotunnel.* TunnelTalk, February 2011

Robbins (2021), *The Channel Tunnel.* Therobbinscompany.com, 2021

Russel et al. (2019), *Transformative Tunnel.* ASCE, Civil Engineering, December 2019

Schmid R, Wagner D (2020), *Eurasia Tunnel Istanbul – Sealing Injections Against High Water Pressure.* Tunnel magazine, June 2020

Shen WQ et al. (2008), *Fire Evacuation and Rescue Design of Shanghai Yangtze River Tunnel.* Shanghai Yangtze

Toris, J L (2007), *A86-West Paris Ring Road. A Bi-mode TBM.* Underground Space, the 4th Dimension of Metropolises, Taylor and Francis, London, UK, 2007

VINCI (2012), *Tunnel Duplex A86 – A Private Initiative.* International Seminar Long Tunnels: Challenges for Design, Construction and Operation, Paris, France, October 2012

Wallis, S (2011), *Alaskan Way a Done Deal, CM Proposals Called.* TunnelTalk, January 2011

Wallis, P (2012), *Mega Dimensions of Russian Orlovski Project.* TunnelCast, September 2012

Wikipedia (2021), *Channel Tunnel.* Wikipedia, 2021

Wikipedia (2021a), *Gotthard Base Tunnel.* Wikipedia, 2021

Wikipedia (2021b), *Western Scheld Tunnel.* Wikipedia, 2021

Wikipedia (2021c), *Autopista de Circunvalación M30.* Wikipedia, 2021

Wikipedia (2021d), *State Route 99 Tunnel.* Wikipedia, 2021

WSP (2021), *Eurasia Tunnel – Istanbul Strait Road Tube Crossing.* WSP, 2021

Chapter 5

The TISB, TMG and TMF concepts

Abstract

This chapter is about the application of the TISB, TMG and TMF concepts to the construction of TBM traffic tunnels.

The main challenges faced by TBM traffic tunnels are presented, with a focus on improving structural safety when in soft ground in seismic areas, the improvement of safety during operation and reduction construction costs.

The TISB, TMG and TMF concepts, developed to face these challenges, are presented, as well as the most relevant technical issues related to the conceptual design and construction of TBM tunnels based on these concepts, namely, configuration of the tunnels cross-section, structural verifications, technical systems and safety in operation.

5.1 INTRODUCTION

As seen in Chapter 4, with the presentation of notable examples of railway and road tunnels recently built around the world, great progress has been made in recent time in the construction of TBM traffic tunnels. However, shortcomings have also been mentioned, therefore, the construction of TBM traffic tunnels still faces significant challenges.

Important issues are, in particular, reliable solutions to ensure the structural safety of tunnels built on soft ground in seismic areas, the appropriate tunnel concept (number of tubes, cross-sectional area of the traffic galleries, etc.), the most appropriate technical systems (ventilation, fire fighting, etc.), and the appropriate concept of safety in operation.

5.2 MAIN CHALLENGES IN TBM TRAFFIC TUNNELS

5.2.1 TBM tunnels in soft ground in seismic areas

As is known, as in TBM tunnels connections between the precast segments have no tensile strength, ductility is low (they are a kind of "LEGO"), therefore, they are suitable for stiff soil (rock, stiff clay, compacted sand, etc.), in non-seismic areas, being stability essentially ensured by the surrounding ground (the precast segments work mainly as a finish).

In the case of TBM tunnels built in soft soil (mud, soft clay, loose sand, etc.), in seismic areas, their structural safety can be questionable and there is the risk of collapse due to soil movements during earthquakes, unless very thick precast segments are used. Soil treatments, sometimes used to improve seismic behavior, are very expensive and sometimes do not guarantee structural reliability. In addition, in the case of tunnels formed by two or three tubes connected by transverse tubes that act as cross-passages between them, the relative displacements between the main tunnel tubes during earthquakes can risk the collapse of the connections to the transverse tubes.

To overcome these limitations, the TISB concept, for TBM tunnels in soft soil in seismic areas, was recently developed by the author.

5.2.2   Tunnel concepts and safety in operation of TBM traffic tunnels

In conventional TBM tunnels, after the execution of the external wall, a filling is placed at the base of the tunnel, creating a platform for the circulation of the vehicles (trains in the case of the railway tunnels, and cars and trucks in the case of the road tunnels), with the two directions of traffic placed side by side.

With regard to TBM railway tunnels, to meet international safety requirements (EU 2008; UIC 2003), the installation of the two directions of traffic, side by side, in the same tube is only possible in short tunnels. In tunnels over 1 km long, due to the piston effect of the trains, this requires limiting the speed of the trains. In addition, it requires the adoption of complex ventilation systems, to control smoke flow inside the tunnel, in the event of a fire.

Thus, as mentioned in Chapter 4, it became common to use the **dual-tube tunnel concept**, in which two separate tunnels are built, each for a direction of traffic, and a system of cross-passages, regularly spaced, interconnecting the two tubes, for access of the emergency personel and rescue of users in incident situations. In these situations, users will leave the incident train and walk to the other railway gallery through the cross-passages, to be later rescued by another train. In very long tunnels, in addition to the cross-passages, large caverns are also usually created, to function as "safe places" for users to await rescue in an incident situation, as well as cross-overs to allow trains to pass from one track to the other. These are the cases of the Guadarrama tunnel, in Spain, and the Gotthard-Base tunnel, in Switzerland, also presented in Chapter 4.

Sometimes, particularly in long underwater tunnels, the **three-tube tunnel concept** is used, in which, in addition to the two railway tubes, a third tube is adopted, placed between the other two, to serve as a service gallery for the evacuation of users and the access of the emergency services in an incident situation, the three being connected by cross-passages. Large caverns are also usually created, namely to install cross-overs to allow trains to pass from one track to the other. This is the case of the Channel Tunnel, between the United Kingdom and France, also presented in Chapter 4, as well as the Basic lay-out of the Gibraltar Strait tunnel, planned to be built between Spain and Marocco (figure 5.1).

However, the **mono-tube dual-gallery tunnel concept** was also used, in which a single large diameter TBM tube, provided with a separating wall in the middle, houses two independent and isolated railway galleries, placed next to each other, as it is the case of the Goeneheart Tunnel, in The Netherlands, also presented in Chapter 4. In the event of an accident or fire inside one of the railway galleries, users will leave the incident railway gallery to the adjacent railway gallery through escape doors regularly arranged in the separating wall, to be rescued later by another train.

With regard to TBM road tunnels, the installation of the two directions of traffic in the same tube is also possible only in short tunnels and with a single lane in each direction.

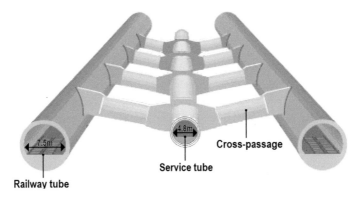

Figure 5.1.   Basic lay-out of the Gibraltar strait tunnel

When there are two or more lanes in each direction, the required diameter would become so large that it would be impractical. Furthermore, in tunnels longer than 0.5 km, to meet safety requirements: ventilation, fire fighting, etc. (EU 2004), placing bidirectional traffic side by side in the same tube becomes very complex and quite expensive.

Hence, the **dual-tube tunnel concept** is now commonly used, in which the tunnel is formed of two separate tubes, each for a direction of traffic, the two being interconnected by cross-passages so that in the event of an incident situation, users will leave the incident tube and move to the adjacent tube, where they will later be evacuated by conventional buses. In Chapter 4, noteworthy examples were presented, the Westerschelde in The Netherlands, the M30 South By-Pass in Madrid, Spain, and the Changjiang in Shanghai, China.

However, another concept came into use, the **mono-tube double-deck tunnel concept**, in which a single large diameter tunnel houses two superimposed road galleries, each for a traffic direction, which are connected together or to outside by systems of evacuation galleries. Noteworthy examples, with two lanes in each direction, were built and presented in Chapter 4: the A86 Duplex in Paris, France, the Eurasia in Instambul, Turkey, and the SR99 in Seattle, USA. A solution with three lanes in each direction, the Orlovski Tunnel, in Saint Petersburg, Russia, was also presented, although it was not built.

In summary, some conclusions can be drawn.

Regarding the environmental impact, tunnels made up of a single tube have less impact than those made of two or three tubes, as, for example, the access ramps will require less space. In addition, when built in urban areas, the impact at surface will also be much less.

As for costs, although the cross-sectional areas of the mono-tube tunnels and the corresponding dual-tube tunnels are similar, in the latter it is necessary to build a complex system of cross-passages, so the cost will inevitably be higher.

With regard to safety in operation, although in those tunnels the most advanced existing safety concepts were adopted, they are not entirely satisfactory, as seen in some events in the recent past.

In order to overcome the above mentioned limitations, the TMG and TMF concepts for TBM railway and road tunnels, respectively, were recently developed by the author.

5.3   THE TISB, TMG AND TMF CONCEPTS

5.3.1   The TISB concept

The TISB ("Tunnel of Improved Seismic Behavior") concept is an innovative solution for TBM tunnels, when the referred tunnels are built in soft ground, in seismic areas, allowing the tunnel be provided with adequate resistance and ductility, without a very significant increase in construction costs. It will also allows for the strengthening of existing TBM tunnels, using them as external formwork for the execution of the interior strengthening.

The TISB concept is a Portuguese Patent (INPI 2008) and is illustrated in figure 5.2 (Pompeu-Santos 2012, 2014, 2016, 2020).

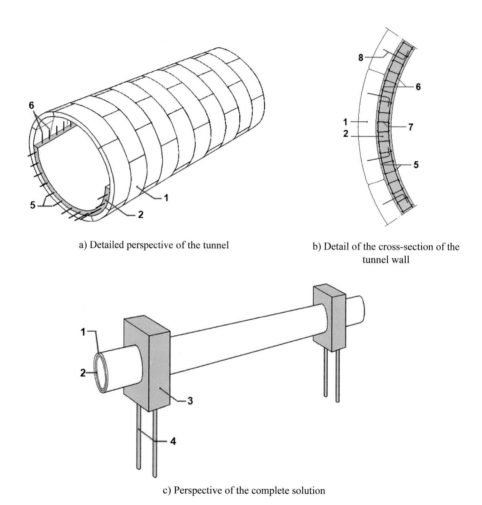

a) Detailed perspective of the tunnel

b) Detail of the cross-section of the tunnel wall

c) Perspective of the complete solution

Figure 5.2.   Illustration of the TISB concept

In the TISB concept, the tunnel is formed by two concentric tubes; an outer tube (1), which is a conventional TBM tunnel, and an inner tube (2), which is subsequently executed, inside the outer one. The outer tube (1) is thus formed by precast segments mounted by the TBM, while the inner tube (2) is later cast, using the latter as exterior formwork.

Within the thickness of the inner tube (2) longitudinal reinforcement bars (5) and transverse reinforcement bars (6) are laid, both in two layers, which are confined by confinement bars (7), so as to provide the tunnel with adequate strength and ductility. Reinforcement bars (8) protruded from the precast segments connect the two tubes.

Where the tunnel can be subjected to significant vertical displacements (due to an increase or decrease in the height of the overburden on weak soil, for example), the tunnel will be provided with supports, regularly spaced along the tunnel axis.

The supports are composed of groups of piles (4) with great horizontal deformability and ductility, arranged in the longitudinal and transverse directions, which are anchored at the top in large blocks of jet-grouting (3) surrounding the outer tube (1), and at the base in the stiff ground below, so to resist vertical loads, while allowing horizontal movements of the tunnel during earthquakes, functioning as a kind of "movable bearings".

The TISB concept thus leads to the obtaining of monolithic structures (joints are eliminated) with appropriate resistance in both longitudinal and transverse directions and great ductility under earthquakes. It will also be very effective if liquefaction and cyclic mobility phenomena occur. In addition, the tunnels will have great structural redundancy (robustness), which can be useful in the case of scenarios not foreseen in the design phase.

5.3.2 The TMG concept

The TMG ("Tunnel Multi Gallery") concept, for TBM railway tunnels, allows, with a suitable arrangement of the tunnel cross-section, that a single-tube tunnel provides the capacities there are normally offered by two or three parallel tubes, thus reducing construction costs, as well as the adoption of a very reliable safety concept, thus significantly improving safety in operation.

The TMG concept allows in a mono-tube TBM tunnel the creation of two independent and isolated railway galleries and a service (emergency) gallery at bottom, inside which appropriate means are installed that create a dedicated and very reliable system for local access of the emergency personnel and evacuation of users to out of the tunnel in the event of an accident or fire inside the tunnel.

The TMG concept is a Portuguese Patent (INPI 2009a) and is illustrated in figure 5.3 (Pompeu-Santos 2012, 2014, 2016, 2020).

In the TMG concept the tunnel is constituted by the outer wall (1) made by the TBM, a slab (3), placed slightly above the bottom of the tunnel and the entire width, and a separating wall (2), placed in the middle of the tunnel and its entire height, so as to form two independent rail galleries, disposed side by side (4) (5), one for each track, and a service gallery (6) below.

In both sides of the tunnel, vertical access galleries (7), regularly spaced and provided with escape doors (8) are created, allowing for the safe passage of people between the railway galleries and the service gallery (6), in the event of an accident or fire inside the tunnel.

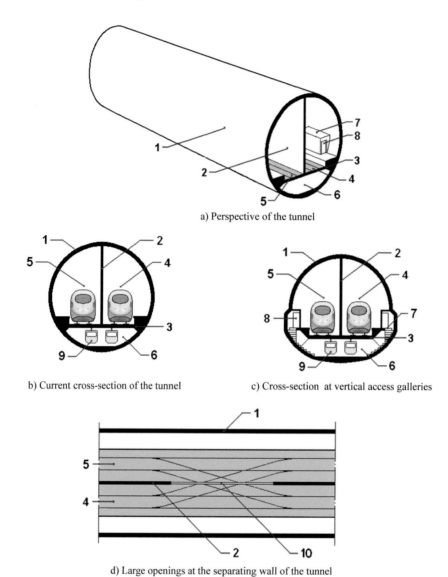

a) Perspective of the tunnel

b) Current cross-section of the tunnel

c) Cross-section at vertical access galleries

d) Large openings at the separating wall of the tunnel

Figure 5.3.   Illustration of the TMG concept

Inside the service gallery (6), emergency vehicles (9) of monorail type are installed, which are parked at the tunnel ends, to provide local access for emergency personnel and the evacuation of users out of the tunnel.

In long tunnels large openings are created, at adequate distances, in the separating wall, provided with sliding fire doors (10), to allow trains to pass from one railway gallery to the other, when it will be necessary to isolate a section of a railway gallery.

A variant B of the basic TMG concept may also be adopted, in which the vertical access galleries (7), instead of being installed on the exterior side of the railway galleries, are placed in the middle of the tunnel, in the separating wall.

Figure 5.4 shows variant B of the TMG concept.

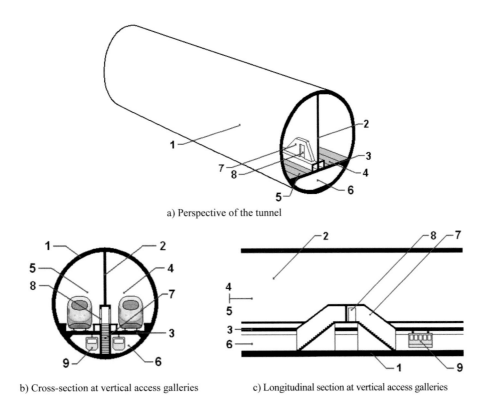

a) Perspective of the tunnel

b) Cross-section at vertical access galleries   c) Longitudinal section at vertical access galleries

Figure 5.4. Illustration of the TMG concept – Variant B

Although in variant B there is a local slight reduction of the cross-section of the railway galleries, it is avoided the need to make openings in the outer wall of the tunnel.

5.3.3 The TMF concept

The TMF ("Tunnel Multi Floor") concept, for TBM road tunnels, allows, with a suitable arrangement of the tunnel cross-section, that a single-tube tunnel provides the capabilities there are normally offered by two parallel tubes, thus reducing construction costs, as well as the adoption of a very reliable safety concept, thus, significantly improving safety in operation.

The TMF concept allows in a mono-tube TBM tunnel the creation of two independent and isolated road galleries and a service (emergency) gallery at the bottom, inside which appropriate means are installed that create a dedicated and very reliable system for local access of the emergency personnel and evacuation of users to out of the tunnel, in the event of an accident or fire inside the tunnel.

The TMF concept is a Portuguese Patent (INPI 2009b) and an European Patent (EPO 2011) and is illustrated in figure 5.5 (Pompeu-Santos 2012, 2014, 2016, 2020).

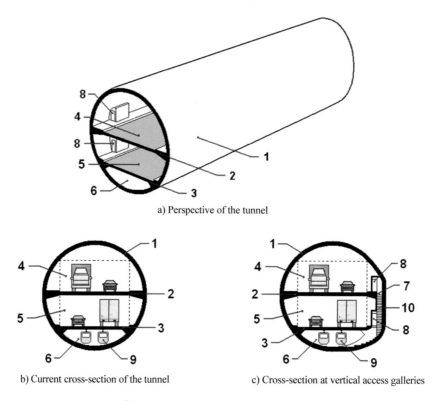

a) Perspective of the tunnel

b) Current cross-section of the tunnel   c) Cross-section at vertical access galleries

Figure 5.5.   Illustration of the TMF concept

In the TMF concept the tunnel is constituted by the outer wall (1) made by the TBM and two slabs (2) (3), built at its full width, one placed roughly at half the height of the tunnel and the other placed slightly over the bottom of the tunnel, so as to form two super-imposed road galleries (4) (5), one for each direction of traffic, and a service gallery (6) below.

In one of the sides of the tunnel, vertical access galleries (7), regularly spaced and provided with escape doors (8) in both road galleries, are also created, allowing for the safe passage of people to the service gallery (6), in the event of an accident or fire inside the tunnel.

Inside the service gallery (6), emergency vehicles (9) of monorail type are installed, which are parked at the tunnel ends, to provide local access for emergency personnel and the evacuation of users to out of the tunnel.

5.4 TUNNELLING BASED ON TISB, TMG AND TMF CONCEPTS

5.4.1 Tunnel cross-sectional layout

The lay-out of the current cross-sections of the railway (double track) tunnels and of road tunnels (with two lanes in each direction) based on the TMG and TMF concepts, respectively, are presented on figures 5.6 and 5.7.

As regard to railway tunnels two alternatives solutions are considered: Alternative A- Tunnels based on Basic TMG concept and Alternative B- Tunnels based on Variant B of the TMG concept.

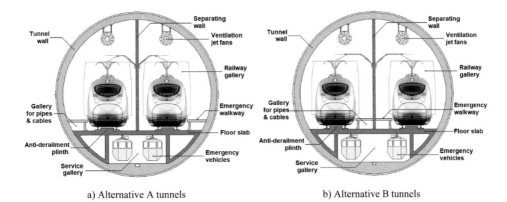

a) Alternative A tunnels   b) Alternative B tunnels

Figure 5.6. Current cross-sectional layout of railway tunnels based on the TMG concept

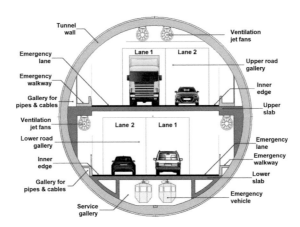

Figure 5.7. Current cross-sectional layout of road tunnels based on the TMF concept

Figures 5.8 and 5.9 show the lay-out of the cross-sections of the tunnels at the vertical access galleries.

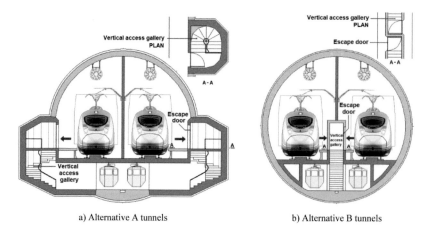

a) Alternative A tunnels   b) Alternative B tunnels

Figure 5.8.   Cross-sectional layout in the areas of vertical access galleries of railway tunnels based on the TMG concept

As regard to road tunnels, vertical access galleries can be placed perpendicular or parallel to the tunnel axis (figure 5.9).

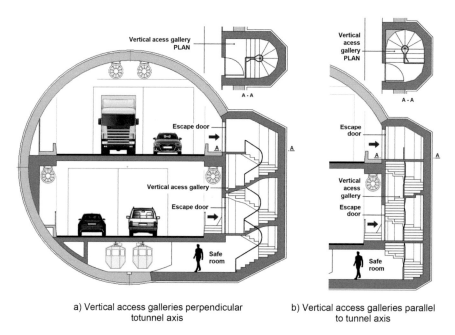

a) Vertical access galleries perpendicular   b) Vertical access galleries parallel
totunnel axis   to tunnel axis

Figure 5.9.   Cross-sectional layout in the areas of vertical access galleries of road tunnels based on the TMF concept

It is worth noting, both in railway tunnels and in road tunnels, an excellent use of the tunnel cross section, leading, in the case of road tunnels, to high values of the respective efficiency index.

It should also to be noted that in the road tunnels, the longitudinal ventilation fans in the lower traffic gallery are located over the emergency walkways, so they not occupy space on the height of that gallery.

5.4.2 Inner diameter of tunnels

Railway tunnels
As mentioned in Chapter 3, in TBM railway tunnels, the inner diameter depend basically on the speed of the trains and the permissible pressure variation inside the trains.

Figure 5.10 shows relationships between the cross-sectional area of single-track railway galleries and the pressure variation inside the trains, for trains of 12.4 m^2 of front area (a common value in high-speed trains), for different train speeds (Tielkes 2006).

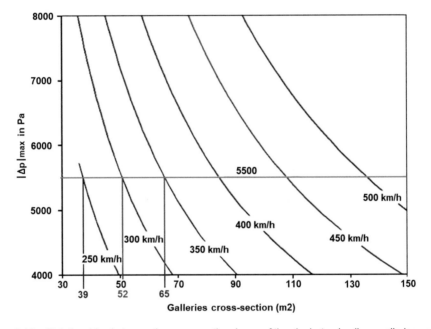

Figure 5.10.   Relationships between the cross-sectional area of the single-track railway galleries and the pressure variation inside the trains, for different train speeds

It can be seen that the speed of the trains is only relevant for tunnels in high-speed railway lines (HSRL), that is, tunnels in lines for trains with speeds from 250 km/h.

Admitting a pressure variation inside the trains of 5.5 kPa (appropriate value in the case of single track tunnels, as there is no clash of the piston effect of the trains), as it can be seen, for speeds of 250 km/h, 300 km/h and 350 km/h, cross-sectional areas of about 39 m^2, 52 m^2 and 65 m^2, respectively, are required in each railway gallery.

However, as mentioned in Chapter 3, this effect is sensitive essentially in the portal areas of the tunnel (at the entrance and exit of the trains), so in tunnels based on TMG concept it will be overcome by adopting appropriate measures in these areas: creation of openings in the separating wall to act as pressure relief, whose area decreases from the outside to the inside, and the adoption of edges inclined about 35° in relation to the horizontal plane (figure 5.11), which allow a reduction of 15 to 20% of the cross-sectional area of the railway galleries (Bogaert 2009).

Longitudinal section   **Cross-section**

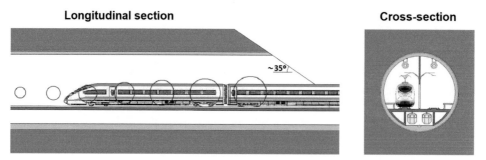

Figure 5.11.   Openings in the separating wall and sloping edges in the portal areas of the HSRL tunnels to reduce the pressure variation inside the trains

Of course, for this solution to work, the entry and exit of the trains in the railway galleries cannot coincide. Now, as these measures cover lengths less than 80 meters, the entry and exit of trains in the tunnel shall be phased out about one second, which, obviously, will not be difficult to guarantee.

When Variant B of the TMG concept is used, the placement of the vertical access galleries in the middle of the tunnel causes a slight decrease in the cross-sectional area of the railway galleries in these areas. However, as the vertical access galleries are located inside the tunnel, outside the portal areas, their presence will have no significant influence on the performance of the solution.

As, it can be seen, this solution of pressure relieves is very cost-effective, compared to those used in the Channel Tunnel or even in the Groeneheart Tunnel, for example (see, Chapter 4). Construction costs are much lower and the openings do not need to be closed in the event of a fire inside the tunnel. It should also be noted that in this solution there is no need to extend the length of the tunnel in the ends, as is necessary in the dual-tube tunnels, as in the Perthus Tunnel (Bogaert 2009) (see, Chapter 3).

In conclusion, in tunnels based on the TMG concept, the separating wall placed in the middle of the tunnel allows to be used for the easy installation of pressure relieves at the portal areas, which allow a significant reduction in the cross-sectional area of the traffic galleries, with practically no additional costs.

Road tunnels

As regard to road tunnels based on TMF concept, the inner diameter will depend essentially on the number of lanes in each road gallery and their width, and the height allowed for vehicles.

In Europe, where the height of the vehicles is limited to 4.0 meters (EU 1996), a minimum clearance of 4.30 meters in road galleries have to be adopted, to take into account a comfort margin to absorb vertical movements. However, if traffic inside the tunnel be limited to light vehicles, a lower height may be adopted (see, Chapter 4).

As is shown in the following chapters, the TMF concept achieves an optimized use of the tunnel cross-section, allowing the obtainment of very cost-effective TBM road tunnels, with efficiency indexes much higher than those of the tunnels built so far, namely of those presented in Chapter 4.

5.4.3 Access ramps of road tunnels

As in the TBM road tunnels based on the TMF concept the two traffic galleries are super-imposed, the access ramps at the ends need an adequate geometry, in order to carry out the transition between these galleries and the adjoining motorway lanes, on which the road platforms are placed side by side.

It can, in general, be assumed that the access ramps have gradients of 5% and that the gap to be overcome is about 5.50m, so the curves in plan have approximately 900m in radius.

Two alternative solutions can be adopted: Alternative A, in which the tunnel and the motor-way are kept aligned and a curve and a counter-curve are adopted (figure 5.12), and Alternative B, in which a single curve is adopted (figure 5.13).

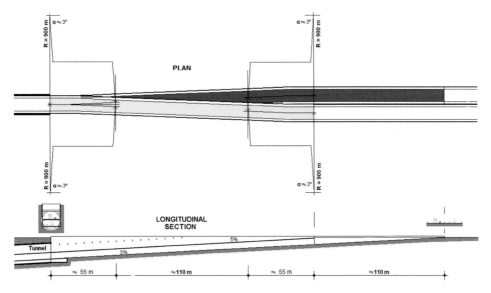

Figure 5.12.   Access ramps of the road tunnels. Alternative A solution

In Alternative A solution, the angle of the curve and the counter-curve will be approxi-mately 3°. The curved sections will be 55m long and connected by a straight section about 110m long.

In Alternative B solution, the angle of the curve will be of approximately 6°. The transitions will consist of a straight section and a curved section, each about 110 meters long.

In the lower gallery, in addition to the aforementioned sections, there will be an additional section, in a straight line, whose length will be greater or less than 110 meters, depending on the gradient of the motorway in the transition zone.

When the tunnel is horizontal in the portals, it is only necessary to make one of the galleries reach the level of the other. In plan, the solution layout will be one of the previous ones, while the longitudinal profile is as shown in figure 5.14.

Despite some technical complexity, these special access ramps will not represent a significant increase in the construction costs of the tunnels.

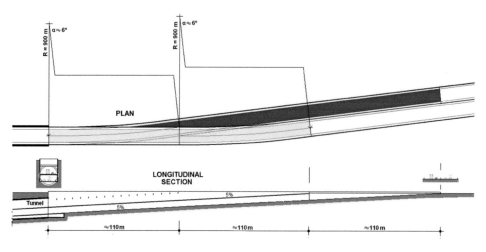

Figure 5.13.   Access ramps of the road tunnels. Alternative B solution

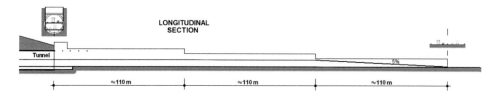

Figure 5.14.   Connection ramp between gallery levels in road tunnels horizontal in the portals

5.4.4   Structural verifications

Tunnels based on the TISB, TMG and TMF concepts behave and are analysed like any other TBM tunnel, but taking into account their specific constitution.

So, these tunnels are analysed as composite structures, considering the precast segments and the cast-in-situ elements working jointly (as composite members) to face the acting actions, both the permanent (self-weight and overburden, earth pressure, water pressure, etc.) and the seismic actions, taking into account the ground-structure interaction. The effect of possible misalignments and excessive rotations between precast segments must also be considered.

To simulate the ground-structure interaction, numerical modelling methods are generally used. These include Finite Element Method (FEM), Finite Difference Method (FDM), and Boundary Element Method (BEM). Continuous analysis is generally adopted, where the domain can reasonably be assumed to be a homogeneous medium (FHWA 2009).

Two-dimensional analysis is, in general, performed. Figure 5.15 shows the schematic structural models of the railway and road tunnel structures, respectively.

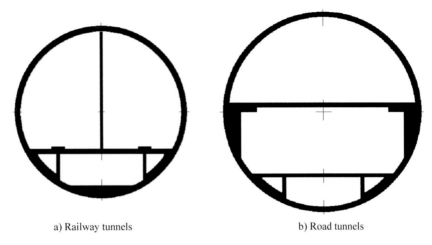

a) Railway tunnels   b) Road tunnels

Figure 5.15.   Schematic structural models

In particular situations, such as simulation of the excavation process or execution of openings in the lining, 3D analysis will be useful. Commercial software that takes into account the actual geological conditions and tunnel construction procedures is available.

The connections between the structural elements (precast segments and cast in-situ elements) are generally considered to be continuity connections. In railway tunnels, if in specific cases it is advantageous, the connection at the upper end of the separation wall will be free to move vertically.

The relevant codes will be applied, for example, Eurocode 2 (CEN 2004b, 2006), Eurocode 7 (CEN 2004a) and Eurocode 8 (CEN 2004c), in the EU and other countries.

From the knowledge of the geologic-geotechnical conditions along the tunnel alignment, a ground profile type or several profiles corresponding to sections with homogeneous characteristics will be defined and the response of the structure in each of these stretches will be determined.

Regarding the seismic behavior, analysis usually considers three primary modes of deformation: 1) cross-section ovalization, 2) curvature deformations and 3) axial deformations.

The following limit states shall be checked: Damage limitation state, which is associated with the functionality of the tunnel (rails alignment, water tightness, etc.) and the Ultimate limit state, which is associated with the collapse of the tunnel lining.

Under the Ultimate limit state, the criteria is the limitation of the strain values on concrete. According to Eurocode 8 (CEN 2004c), in unconfined concrete strains are limited to 0.0035 while in confined concrete they can be 0.0100 or even higher, provided that the strains are predominantly due to flexure.

To ensure the seismic safety the tunnels, two different strategies can be adopted.

One possibility is to increase the thickness of the tunnel wall (lining), however, in areas of high seismicity and poor soft ground, the required thickness may not be realistic.

Alternatively, an approach based on the TISB concept may be used, that is, the adoption of additional properly confined reinforced concrete elements cast in-situ, in order to create a ductile inner tube, acting jointly with the TBM tube (figure 5.16). Concrete elements already needed inside the TBM tube, will form part of the inner tube.

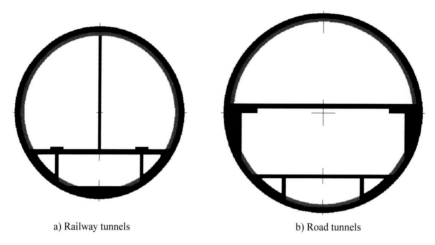

a) Railway tunnels   b) Road tunnels

Figure 5.16.   Ensuring the seismic safety of traffic tunnels with the use of the TISB concept

In addition to reducing acting stresses, it allows a great increase in the values of the allowable strains in concrete of the inner tube and thus on the inelastic deformations of the tunnel.

The thickness of the precast segments will thus be determined by the actions acting during the construction phase of the tunnel, in which the seismic action will be significantly reduced.

It should also be noted that the cast-in-situ slabs, connected in continuity with the precast segments, as well as the other additional reinforced concrete elements cast inside the tunnels, will significantly improve the static and seismic behaviour of the tunnels on soft ground.

In fact, these elements will significantly stiffen the tunnel for horizontal actions, namely acting as horizontal ties, as well as reducing the height of the tunnel's cross-section that will be deformed horizontally under earthquakes, which will be essentially limited to the upper vault of the tunnel wall.

5.4.5   Execution of traffic galleries inside the tunnels

In both railway and road tunnels, the execution of the traffic galleries inside the TBM tunnel will be carried out according to the progress of the construction of the tunnel, although with some hundred meters delay, so that the works do not conflict with the operation of the TBM.

However some works can only be carried out after the completion of the construction of the tunnel, in order to allow the circulation of vehicles with supplies for the TBM.

Figures 5.17 and 5.18 show the foreseen sequence of the execution stages of the traffic galleries in the railway and road tunnels, respectively.

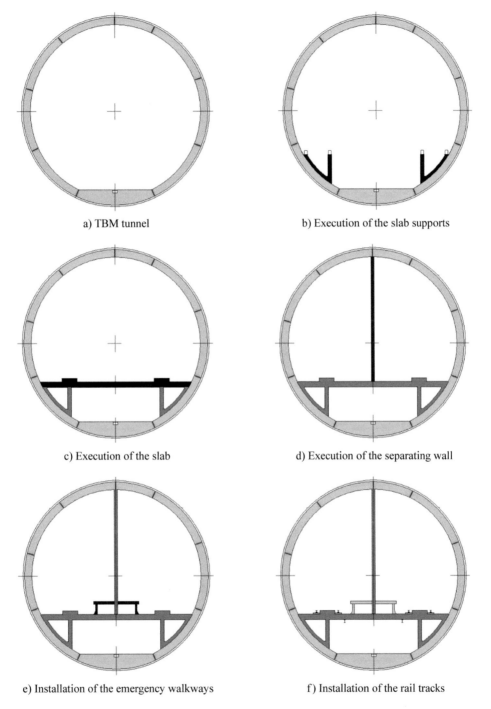

a) TBM tunnel

b) Execution of the slab supports

c) Execution of the slab

d) Execution of the separating wall

e) Installation of the emergency walkways

f) Installation of the rail tracks

Figure 5.17. Sequence of the stages of execution of the traffic galleries of railway tunnels (Alternative B tunnels)

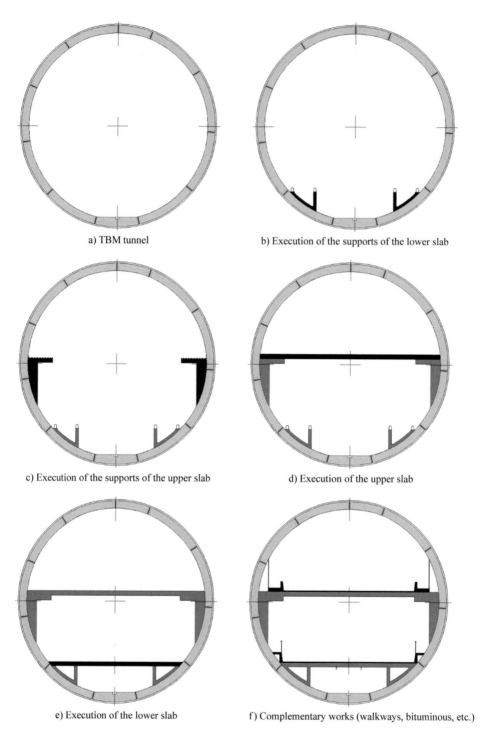

a) TBM tunnel

b) Execution of the supports of the lower slab

c) Execution of the supports of the upper slab

d) Execution of the upper slab

e) Execution of the lower slab

f) Complementary works (walkways, bituminous, etc.)

Figure 5.18.   Sequence of the stages of execution of the traffic galleries of road tunnels

5.4.6 Connections of cast-in-situ elements to the tunnel wall

The concrete elements cast inside the tunnel are connected to the outer wall of the tunnel (tunnel lining) through reinforcement bars protruding from the internal face of the precast segments that assure the structural continuity (figure 5.19) (Wikipedia 2021).

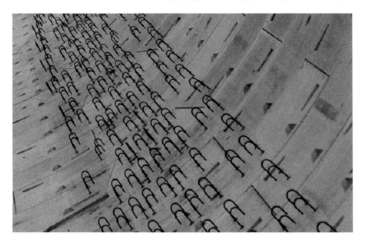

Figure 5.19. Reinforcement bars anchored on the intrados of the precast segments

These bars are anchored in holes previously made on the inner face of the precast segments or drilled when these elements are already in place, which the remaining space subsequently filled with high-strength non-shrinkage mortar.

5.4.7 Execution of vertical access galleries

In the execution of vertical access galleries, two different situations have to be considered.

When these galleries are located inside the tunnel, on the separating wall, case of the Alternative B railway tunnels, their execution do not put any particular difficulty, being carried out in conjunction with the execution of the inside traffic galleries.

When these galleries are located on the outer wall of the tunnel (lining), cases of the road tunnels and the Alternative A railway tunnels, they will be built by locally disassembling special segments (steel boxes) mounted in these areas (belonging to two rings, in principle), excavating the necessary space in the ground aside the tunnel and casting concrete walls in-situ, which are connected in continuity to the adjacent precast segments by inserting reinforcement bars in holes made in their surfaces.

However, it should be noted that the execution of these works is similar to the situations faced in the construction of some tunnels presented in Chapter 4, namely the access galleries to the technical gallery of the railway Groenehart Tunnel, in The Netherlands (see, figure 4.15), or the connections between levels in the M30 South By-Pass Tunnel, in Madrid (Spain) (see, figure 4.21), and the safety recesses and the emergency access shafts in the road Duplex Tunnel, in Paris (France) (see, figures 4.28 and 4.29).

An important task is to guarantee the stability of the tunnel lining when disassembling the special steel segments. As in the construction of openings for cross-passages in TBM dual-tube tunnels, two alternative strategies can be adopted: a) installation of temporary support systems adjacent to the openings (figure 5.20a) (Lee 2017); b) adoption of permanent shear interfaces (shear keys, etc.) between precast segments of adjacent rings (figure 5.20b) (Valle 2014, modified for conceptual purposes). In any case, these measures must be in place before starting to disassemble the segments.

a) Temporary steel frame support system   b) Concrete shear keys between precast segments

Figure 5.20.   Ensuring the stability of the tunnel lining during the execution of vertical access galleries

For the dimensioning of these devices and the local design of the lining, 3D finite element analysis of these tunnel areas will be convenient. It should be noted that (after being built) the walls of the vertical access galleries ensure the structural continuity of the tunnel lining.

If there is no water pressure in the place (or if the water flow is low) the works are carried out excavating directly inside the ground. To stop or to impede any water from entering in the excavation space, injections of cementitious grout or resin foams are usually efficient.

However, when there is water pressure around the tunnel specific additional measures have to be taken. One solution presumed to be effective, is the creation of a kind of "cake" on the site, made of grout injections (jet grouting), inside of which the works are developed.

Thus, the steel segments are provided with holes, to allow for the execution of the injections. The precast segments of the adjacent rings are also provided with holes for making complementary injections, although not being disassembled.

It should be noted that the execution of vertical access galleries on the outer wall of the tunnels when there is water pressure around it, despite some risks, is less complex and expensive than the execution of cross-passages in the dual-tube TBM rail and road tunnels.

In fact, as it can be seen, the vertical access galleries have smaller dimensions and less demanding requirements than the cross-passages in those tunnels, so the cost of their execution will also be less.

In road tunnels, the works will be carried out in three phases, corresponding to the three levels of these galleries: the upper level, the intermediate level and the lower level. In Alternative A railway tunnels the methodology is similar, being carried out in two phases, corresponding to the two levels of these galleries: the upper level and the lower level. The sequence of the execution works of vertical access galleries in road tunnels, when there is water pressure around the tunnel, is illustrated in figure 5.21.

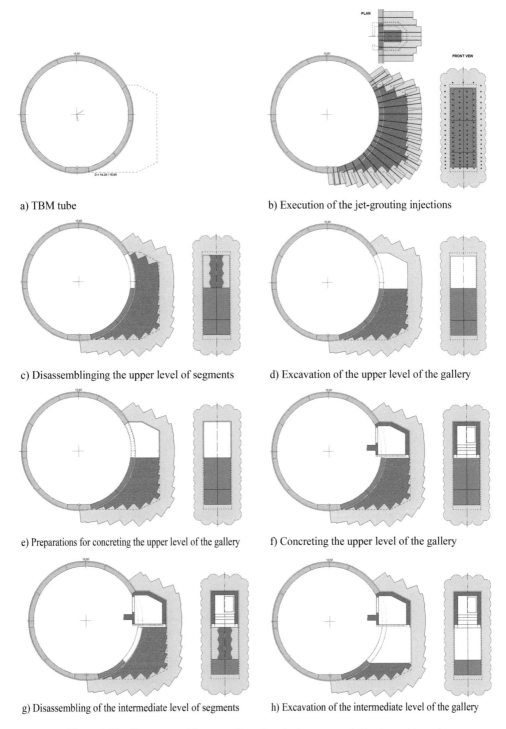

Figure 5.21.   Sequence of the execution of vertical access galleries in road tunnels

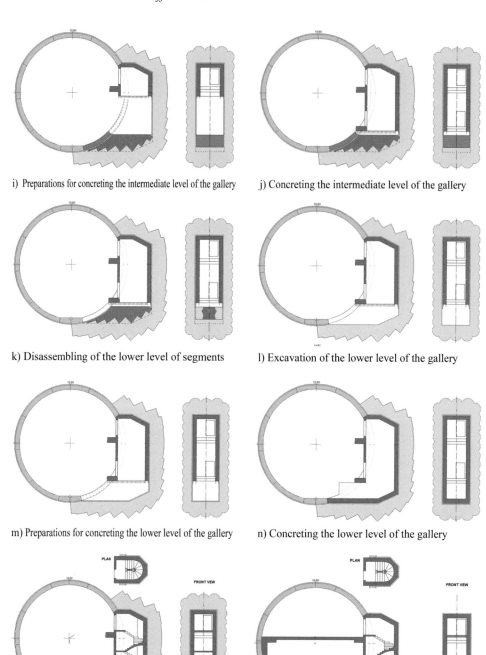

i) Preparations for concreting the intermediate level of the gallery   j) Concreting the intermediate level of the gallery

k) Disassembling of the lower level of segments   l) Excavation of the lower level of the gallery

m) Preparations for concreting the lower level of the gallery   n) Concreting of the lower level of the gallery

o) Concreting of the stairs inside the gallery   p) Concreting of the remaining elements inside the tunnel

Figure 5.21.   Continued

5.4.8 Technical systems

Drainage

Although the seals placed at the joints between precast segments of TBM tunnels are, in general, quite efficient, when there is external water pressure, there always be some water ingress. In addition, although the firefighting systems adopted in both railway and road tunnels, based in the TMG and TMF concepts, respectively, does not require much water, it will be necessary to drain any water that enters them.

Thus, both traffic galleries of these tunnels are provided with drainage systems, consisting of lines of sink drains, regularly distributed along their length, with the effluent being discharged into pipes placed along the underside of the pavement slabs, from which it is conducted for wells that are created along the bottom of the tunnel (figures 5.22 and 5.23).

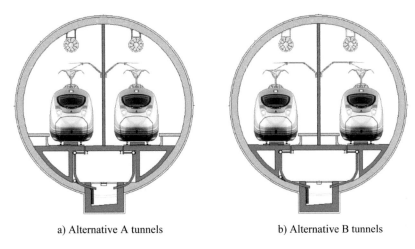

a) Alternative A tunnels   b) Alternative B tunnels

Figure 5.22.   Drainage system in railway tunnels

The service gallery of both railway and road tunnels is also provided with a channel located in the middle, to conduct any receiving water to the wells.

From the wells, the effluent is then pumped out of the tunnels through pipes placed along one of the sides of the service gallery.

The pavements of the traffic galleries have a gradient of 0.5% for the lines of sink drains.

If there are discharges of fuel or other contaminating liquids in the pavements of the traffic galleries of the road tunnels, it will be possible to remove them from the wells.

Fire fighting

In the railway and road tunnels based on the TMG and TMF concepts, the traffic galleries are equipped with active detection systems and WFSs ("Water Based Fire Fighting Systems") (see, Chapter 3).

Heat sensor and smoke detection systems automatically activate high pressure water mist nozzles, distributed along the tunnel and grouped into sections (each about 30 meters long), which will drastically reduce the temperature at the source of fire (figures 5.24 and 5.25).

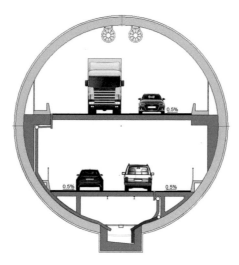

Figure 5.23.   Drainage system in road tunnels

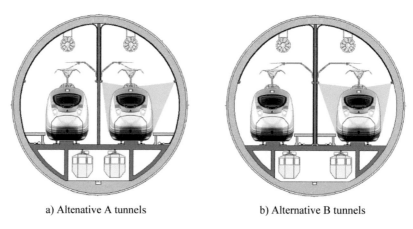

a) Altenative A tunnels   b) Alternative B tunnels

Figure 5.24.   Fire fighting systems in the traffic galleries of railway tunnels

After this action, fire fighters, who access the fire site through the service (emergency) gallery in special emergency vehicles, will then easily extinguish the fire, using local fire hydrants.

With regard to railway tunnels, the nozzles are distributed along the so-called "emergency stations", with a length at least equal to that of the longest trains that will circulate in the tunnel (750 meters, in principle), where appropriate comfort conditions will be available for users (lighting, etc.). Different situations will be considered.

Short tunnels (less than 5 km in length) need not have an "emergency station".

Medium-size tunnels (between 5 km and 10 km long), will have an "emergency station" located at the mid length of the tunnel, being, in the event of a train on fire stopping, activated

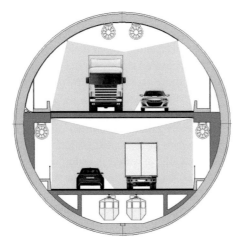

Figure 5.25.   Fire fighting system in the traffic galleries of road tunnels

the nozzles belonging to the sections placed along the length of the burning carriages or wagons, which can significantly reduce the number of activated nozzle sections.

In long tunnels (over 10 km long) there will be "emergency stations" spaced 5 km at most, being, in the event of a fire, activated the nozzles belonging to the sections of the "emergency station" where the train stops, along the length of the burning carriages or wagons.

Thus, a burning train will take a maximum of 4 minutes to reach the outside or the next "emergency station" (average speed of 80 km/h), compared to the 20 minutes required in tunnels such as Guadarrama or Gotthard-Base, without significant cost increase.

With such short distances between "emergency stations", it is virtually guaranteed that there will no trains in fire stopping outside "emergency stations".

With regard to road tunnels, nozzles are distributed along the entire length of the tunnel, being, during a fire, activated the nozzles of three sections: the source of the fire and the adjacent sections.

Water lines, placed inside the galleries under the emergency walkways (in the railway tunnels), or aside the emergency walkway opposite to the escape doors of the upper traffic gallery (in the road tunnels), supply pressurized water to the systems.

These main lines supply secondary lines corresponding to each section and controlled by valves, which in turn feed the section's nozzles.

Hydrants placed at regular intervals along the outer side of the emergency walkways of the traffic galleries allow fire fighters to access the pressurized water network.

As mentioned in Chapter 3, the use of WFS systems make possible to reduce the requirements on concrete protection and the fire rating of mechanical and electrical equipment of the tunnel.

Ventilation
Since both railway and road tunnels are equipped with WFSs, the ventilation systems in the traffic galleries are of the longitudinal type, with jet fans regularly distributed along the length

of the tunnel, which push air in the direction of the traffic (see, Chapter 3), therefore very simple and inexpensive ventilation systems (see, figures 5.6 and 5.7).

In railway tunnels, fans are placed at the top of the rail galleries. In normal operation fans are not activated, being the air renovated simply by the piston effect of the circulating trains. In fire situations, are activated those fans placed downstream the fire, in order to disperse away from the area the smoke provoked by the burning train.

In road tunnels, fans are placed on the top of the upper road gallery and over the emergency walkways of the lower road gallery. In normal operation fans are also not activated, being automatically activated when the accepted pollution parameters are exceeded. In fire situations, fans will be automatically activated, in order to disperse the smoke downstream of the fire source. However, as also mentioned in Chapter 3, in the vicinity of the fire, the fans must be controlled in order to protect the fleeing users.

The service gallery of both railway and road tunnels has its own ventilation system, with ventilators operating at both ends, in order to create an overpressure inside, so that smoke does not enter these galleries neither the vertical access galleries.

5.4.9   Safety in operation

Procedures for safety in operation (safety concept) of tunnels based on TMG and TMF concepts, that is access of emergency services and rescue of users to out of the tunnel, are, on the one hand, quite simple and on the other very reliable.

Both railway and road tunnels are provided with emergency walkways along the full length. In road tunnels they are, in principle, 1.20m wide, being placed in both sides of each traffic gallery. In railway tunnels they are, in principle, 1.40m wide, being placed on the outside of the traffic galleries in Alternative A tunnels or on the inside of the traffic galleries in Alternative B tunnels.

Both railway and road tunnels are also provided with escape doors and vertical access galleries with spacing in the order of 400m, to accede the service (emergency) gallery below.

The service gallery aims to fulfil two objectives: on the one hand, to function as a "safe place" for users in the event of an accident or fire in the tunnel and, on the other hand, to provide access to emergency services to the site and evacuation of users to out of the tunnel.

With regard to road tunnels, in the event of an incident in one of the traffic galleries, the Control Center asks the drivers to stop and to leave their vehicles. Then, users leave the incident traffic gallery by walking through the emergency walkways to the nearest escape door, from which they access the service gallery, downing the stairs of the respective vertical access gallery. Inclined platform lifts, running along the stairs, provide access to people with limited mobility.

As for the railway tunnels, in the event of activation a fire alarm, the train will continue to out of the tunnel, or, at least to the next "emergency station". In any case, when a train stops inside a traffic gallery, users will leave the train using the respective emergency walkway, walking until the nearest escape door, from which they access the service gallery bellow, downing the stairs of the respective vertical access gallery. Inclined platform lifts, running along the stairs, will also provide access to people with limited mobility.

In both railway and road tunnels, within the service gallery, dedicated EMEVs ("Emergency Monorail Electric Vehicles"), which are parked at one or both portals, will be moved to the incident site, circulating suspended from the lower slab in two parallel lines, allowing emergency personnel (rescue personnel and fire fighters) to access the site.

EMEVs are remotely controlled monorail vehicles, about 4.50m long, 1.60m high and 1.40m wide, battery powered and driven by linear induction motors. They are equipped with smart cameras that receive (and transmit) wireless signals from (and to) the Control Center of the tunnel (figure 5.26). They travel at speed of about 60 km/h, that is, they run about 1 km per minute.

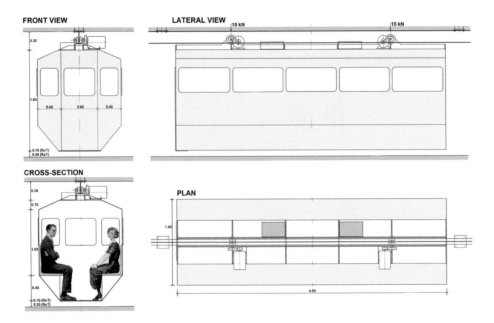

Figure 5.26.   Lay-out of the emergency vehicles (EMEVs)

Thus, while fire fighters will go up to the incident traffic gallery to respond to the situation, the rescue personnel organize the rescue of users to out of the tunnel through the EMEVs. It should be noted that emergency personnel and users never make use of the traffic galleries.

At the tunnel portals (in one or in both), next to the parking of EMEVs, there are access buildings, equipped with stairs and elevators (in principle, located next to the Control Center of the tunnel), where the rescued users will have exit to out of the tunnel.

EMEVs have a capacity of 20 people each (in principle), being grouped in "trains", in numbers according to needs. In road tunnels, the number of units of each "train" will be two (figure 5.27). In railway tunnels, the number of units will depend on the capacity of the passenger trains that will use them. For trains with a capacity of up to 400 passengers, the number of units will be five (figure 5.28).

In both railway and road tunnels less than 10 km long, EMEV "trains" are parked in one of the portals (two EMEV "trains" in total), while in tunnels over than 10 km long there will be EMEV "trains" parked in both portals (four EMEV "trains" in total).

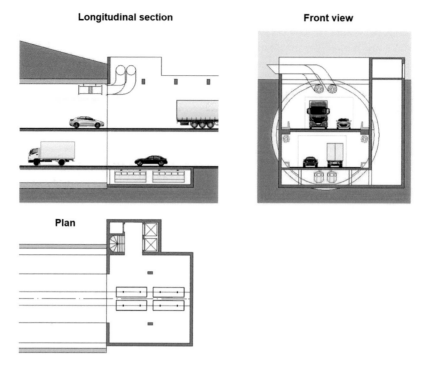

Figure 5.27.   Parking of EMEVs and access building in road tunnels

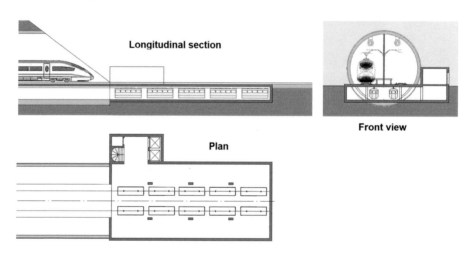

Figure 5.28.   Parking of EMEVs and access building in railway tunnels

In railway tunnels, to rescue passengers from a train with a capacity for 400 passengers, for example, EMEV "trains" will need to make one or two trips each, depending on whether they are four or two, respectively.

In road tunnels, as the number of users to be rescued will, in principle, be relatively low, the EMEV "trains" do not need to make more than one trip each.

In very long railway tunnels (over 40 km long), the four EMEV "trains", although two are parked at each portal, in incident situations the four may park at just one of the portals (the closest to the site of the incident), in order to reduce the rescue time of all users (figure 5.29).

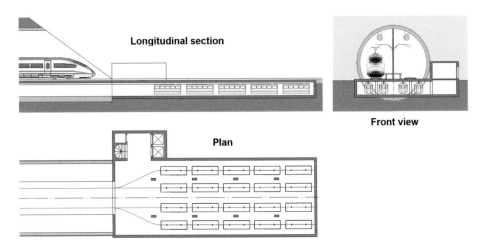

Figure 5.29. Parking of EMEVs in a rescue operation in very long railway tunnels

Railway tunnels over 10 km in length will still be provided with cross-overs, to allow trains to pass from one track to the other, in the case there is the need to deactivate a section of a track (maintenance works, broken train, etc.), which will be spaced 10 km at maximum.

Thus, in these areas, the separating wall will be provided with large openings (about 40m in length and 6m high), protected with fireproof sliding doors (made of several sheets), which will be opened to allow trains to move from one gallery to the other, in case of need.

As seen, the rescue concepts in both the railway and road tunnels are very reliable, being completely independent on traffic conditions in the traffic galleries of the tunnel. In addition, as the emergency services does not make use of the traffic galleries of the tunnel, there will be no disturbance of the traffic flow in the non-incident gallery.

5.5 CONCLUSIONS

Despite the great progress made in recent time, the construction of TBM traffic tunnels still faces significant challenges. Important issues are, inter alia, appropriate measures to ensure the reliability of the structural safety of TBM tunnels built on soft ground in seismic areas, and the most suitable tunnel concept in TBM railway and road tunnels (number of tubes, cross-sectional area of the traffic galleries, etc.), as well as the most appropriate technical systems (ventilation, firefighting, etc.) and the respective safety in operation concept, that is, measures for evacuation of users in the event of an accident or fire inside the tunnel.

To overcome these challenges, three innovative concepts were recently developed by the author: the TISB concept for TBM tunnels in soft soil in seismic areas, and the TMG and TMF concepts, for TBM railway and road tunnels, respectively.

The TISB concept allows TBM tunnels built on soft ground in seismic areas be provided with reliable structural safety, without a significant increase in construction costs. The tunnel is formed by two concentric tubes, an outer tube, which is a conventional TBM tunnel, and an inner tube cast in-situ, reinforced with properly confined longitudinal and transverse reinforcements, so as to give the tunnel with adequate strength and ductility.

The TMG and TMF concepts, for railway and road tunnels respectively, allow, with a an appropriate lay-out of the tunnel cross-section and the adoption of suitable additional measures, that mono-tube TBM tunnels have the capacities normally offered by two or three parallel tubes, thus reducing construction costs, in addition to significantly improving safety in operation.

In both cases, three independent galleries are created inside the tunnel: two traffic galleries (next to each other in the case of railway tunnels or superimposed in the case of road tunnels) and a service (emergency) gallery at the bottom, the three being connected by vertical access galleries, allowing the tunnel to be provided with a very reliable safety system for access of the emergency services and evacuation of users in the event of an accident or fire.

With regard to the inner diameter of the railway tunnels, the adoption of appropriate openings in the portal areas of the separating wall, allows a significant reduction in the cross-sectional area of the traffic galleries, in comparison with tunnels based on the three-tube and dual-tube concepts.

With regard to the inner diameter of the road tunnels, there is an optimized use of the tunnel's cross section, leading to significantly higher values of the respective efficiency index, in comparison with that of tunnels based on the dual-tube and conventional mono-tube concepts.

For the execution of vertical access galleries, suitable technical solutions are available. When they are located on the outer wall of the tunnel and there is water pressure around it, although there are some risks, their construction is no more complex and expensive than building the cross-passages in the dual-tube tunnels.

Tunnels based on the above mentioned concepts behave and are analysed as composite structures, considering the precast segments and the cast-in-situ elements working together to face the acting actions, taking into account the soil-structure interaction.

The fire fighting systems in the traffic galleries of both railway and road tunnels are based on WFSs ("Water Based Fire Fighting Systems"), consisting of high pressure water mist nozzles, distributed along the traffic galleries. In the railway tunnels, nozzles are distributed along the so-called "emergency stations", spaced about 5 km away, while in the road tunnels, nozzles are distributed along the entire length of the tunnel, however in both cases activated only those in the vicinity of the fire.

The ventilation systems of the traffic galleries in both railway and road tunnels are of the longitudinal type, with jet fans, regularly distributed along the tunnel, which push the air and smoke in the direction of the traffic.

Safety in operation (safety concept) of both railway and road tunnels makes use of an innovative concept, quite simple and very reliable (more reliable than any other existing), based on a service gallery, so it is totally independent on traffic conditions inside the tunnels.

The service gallery serves, on the one hand, as a "safe place" for users in an incident situation, and, on the other, to provide access to emergency services and evacuation of users to out of the tunnel, by means of EMEVs ("Emergency Monorail Electric Vehicles").

EMEVs have a capacity of 20 people each (in principle), being grouped in "trains" in numbers according to needs. EMEV "trains" are parked in one or both the portals, depending on the tunnel length, and circulate inside the service gallery, suspended from the lower slab. In addition, as the emergency services and users do not use the traffic galleries, their access will not be conditioned, nor there will be disturbance of the traffic flow in the non-incident gallery.

In summary, the TISB, TMG and TMF concepts are innovative solutions that can be a great asset in the construction of tunnels made using the TBM technique. In the following chapters, examples of application of these concepts are presented.

REFERENCES

Bogaert P (2009), *Current and End Section Design Requirements for High-Speed Railway Tunnels*. Workshop Tunnels for High Speed Railway, Porto, Portugal, October 2009

Bopp R, Hagenah B (2009), *Aerodynamics, Ventilation and Tunnel Safety for High Speed Rail Tunnels*. Workshop Tunnels for High Speed Railway, Porto, Portugal, October 2009

CEN (2004a), *EN1997-1: Eurocode 7- Geotechnical Design. Part 1: General Rules*. CEN, November 2004

CEN (2004b), *EN1992-1-1: Eurocode 2- Design of Concrete Structures. Part 1-1: General Rules and Rules for Buildings*. CEN, December 2004

CEN (2004c), EN1998-1: *Eurocode 8- Design of Structures for Earthquake Resistance. Part 1: General Rules, Seismic Actions and Rules for Buildings*. CEN, December 2004

CEN (2006), *EN1992-3: Eurocode 2- Design of Concrete Structures. Part 3: Liquid Retaining and Containment Structures*. CEN, June 2006

EU (1996), *Weights and Dimensions*. EU Directive 1996/53CE, July 1996

EU (2004), *Safety in Roadway Tunnels*. EU Directive 2004/54CE, April 2004

EU (2008), *Safety in Railway Tunnel*. EU Decision 2008/163CE; January 2008

EPO (2011), *Tunnel Multi-Storey*. EPO, European Patent EP 2317074A1, November 2011

FHWA (2009), *Technical Manual for Design and Construction of Road Tunnels – Civil Elements*. Federal Highway Administration, U.S. Department of Transportation Publication Nº FHWA-NHI-10-034, December 2009

INPI (2008), *Túnel de Comportamento Sísmico Melhorado (Tunnel of Improved Seismic Behaviour)*. INPI, Portuguese Patent n.103421 (in Portuguese), July 2008

INPI (2009a), *Túnel de Galeria Múltipla (Tunnel Multi-Gallery)*. INPI, Portuguese Patent n. 103769 (in Portuguese), January 2009

INPI (2009b), *Túnel Multi Piso (Tunnel Multi-Floor)*. INPI, Portuguese Patent n. 103748 (in Portuguese), August 2009

Lee T H et al. (2017), *Numerical Analysis of Cross-Passage Opening for TBM Tunnels*. 19th International Conference on Soil Mechanics and Geotechnical Engineering, Seoul, 2017

Pliego J M (2005), *Opening Session – Briefing Document*. UN-ITA Workshop Systematic Ground Probing and Treatment in Mechanised Tunnelling, Madrid, Spain, January 2005.

Pompeu-Santos S (2008a), *Tunnels of Improved Seismic Behaviour*. FIB Symposium Taylor Made Concrete Structures – New Solutions for Our Society, Amsterdam, Netherlands, May 2008

Pompeu-Santos S (2008b), *TISB and TMG Concepts- Application on a Proposal for the Railway Tunnel through the Gibraltar Strait*. IABSE Symposium Improving Infrastructure Worldwide, Chicago, USA, September 2008

Pompeu-Santos S (2012), *TBM Tunnels of Improved Seismic Behaviour- The TISB Concept*. 15th World Conference on Earthquake Engineering (15WCEE), Lisbon, Portugal, September 2012

Pompeu-Santos S (2014), *New Trends in the Tunnels of the Future*. IABSE Symposium Engineering for Progress Nature and People, Madrid, Spain, September 2014

Pompeu-Santos S (2016), *Sustainable TBM Tunnels for Tomorrow*. Second International Conference on Concrete Sustainability (ICCS2016), Madrid, Spain, June 2016

Pompeu-Santos S (2020), *Innovative Concepts in TBM Tunnels*. Tunnel Engineering – Selected Topics (Chapter 7), IntechOpen, UK, April 2020

Tielkes T (2006), *Aerodynamic Aspects of Maglev Systems*. 19th Conference on Magnetically Levitated Systems and Linear Drives, Dresden, Germany, September 2006

UIC (2003), *UIC Code 779-9: Safety in Railway Tunnels*. UIC, 2003

Valle N D et al. (2014), *Adoption of Shear Keys to Provide Stability to TBM Tunnels Segmental Lining During Cross Passages Construction: The Example of Aked Tunnel.*

World Tunnel Congress (WTC2014), Foz de Iguassu, Brazil, May 2014

Wikipedia (2021), *State Route 99 Tunnel*. Wikipedia, 2021

Chapter 6

Rehabilitation of a TBM tunnel on soft ground in a seismic area based on the TISB concept

Abstract

This chapter presents the application of the TISB concept on the repair and seismic strengthening of a tunnel section of the Metro network, in Lisbon, Portugal.

As a result of an incident at the final stage of construction of the tunnel, there was a need for its rehabilitation. During the rehabilitation process, the safety against earthquakes of the solution put out to tender was questioned and the contractor developed an alternative solution based on the TISB concept.

The basic and alternative rehabilitation solutions are presented and compared, particularly with regard to seismic behaviour, resulting in the alternative solution being more advantageous.

6.1 INTRODUCTION

As mentioned in Chapter 5, TBM tunnels are suitable for stiff soil, where structural safety is mainly ensured by the surrounding soil (through the soil-structure interaction), with the tunnel wall essentially acting as a lining.

When in soft ground (landfills, alluvia, etc.), in seismic areas, the behaviour of the TBM tunnels is highly dependent on the geotechnical characteristics of the surrounding soil, with the risk of major deformations and significant damage when subjected to earthquakes.

A tunnel section of the Metro network, in Lisbon, Portugal, built using the TBM technique in a very poor ground, suffered an incident at the final stage of its construction, which resulted in the need for its rehabilitation. During the rehabilitation process, an alternative solution, based on the TISB concept, was developed, with a view to also ensure its safety against earthquakes.

6.2 CHARACTERISTICS OF THE TUNNEL

The section of the Lisbon Metro network, between Baixa-Chiado and Praça do Comércio stations in Lisbon, Portugal, is a conventional TBM metro tunnel with two tracks (figure 6.1), which in the final part runs under the Tagus River (Ferconsult, 2002).

The tunnel has an inner diameter of 8.80m, being formed of precast segments (six segments in each ring) 0.36m thick and 1.20m wide, and grout injections 0.12m thick, resulting in an excavation diameter of 9.76m (figure 6.2).

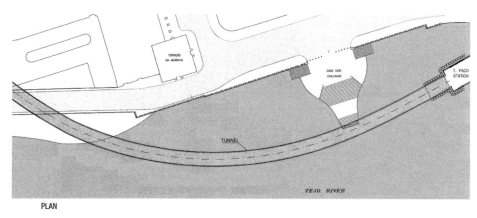

PLAN

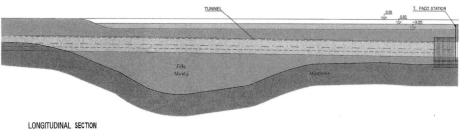

LONGITUDINAL SECTION

Figure 6.1.   Plan and longitudinal section of the tunnel

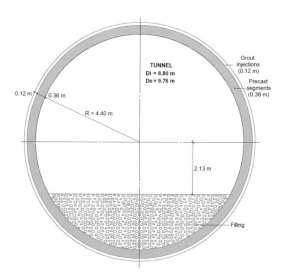

Figure 6.2.   Cross-section of the tunnel

At the base of the tunnel, a filling of porous concrete was placed in order to create a platform for the installation of the metro tracks.

In front of Praça do Comércio, on the section where the tunnel runs under the Tagus River, since the soil overburden is relatively low, a temporary embankment was installed before the passage of the TBM, to facilitate soil compaction and create additional weight to avoid the uplifting of the machine (figure 6.3).

Figure 6.3.   Temporary embankment installed for the passage of the TBM

In this area, the geotechnical conditions of the soil are very poor, consisting of embankments and layers of alluvium and sand that reach more than 40 meters in thickness (figure 6.4).

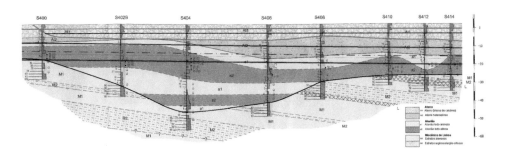

Figure 6.4.   Soil characteristics along the section of the tunnel under the temporary embankment

6.3 THE INCIDENT

At the beginning of the constructing of the Praça do Comércio station, located next to the section of tunnel under the Tagus River, when drilling the tunnel wall to perform jet-grouting injections, an incident occurred: water and soil started to enter the tunnel through the holes drilled on the wall, risking the collapse of the tunnel and of the nearby buildings, of great cultural value.

To avoid the collapse, the tunnel was filled with water in order to equilibrate the water pressure outside and inside the tunnel. With this action, the movement of soil towards the interior of the tunnel was stopped, thus the collapse was avoided. However, with the added weight and weak soil resistance underneath, the tunnel deformed vertically along a length of about 250 meters, in the zone where the layers of alluvia are thicker.

The deformation of the tunnel was accompanied by the opening of a longitudinal crack along its top and, as a result, a significant number of precast segments became dramatically cracked, diminishing the structural safety and tightness of the tunnel.

When the situation was stabilized, works were carried out in the tunnel area where the problem started, to ensure its integrity, and water removed from the interior of the tunnel, to allow the rehabilitation of the structure.

6.4 THE REHABILITATION PROCESS

After several studies carried out by the owner, a tunnel rehabilitation solution, developed by TEC- Tunnel Engineering Consultants, was put up for tender.

In the tender solution the rehabilitation of the tunnel essentially consists of the execution of a new tube, in reinforced concrete, inside the TBM tube. The inner tube is 370m long and is formed by 77 cast insitu segments, each 4.8m long (corresponding to four TBM tube rings), separated by waterproof expansion joints (figure 6.5).

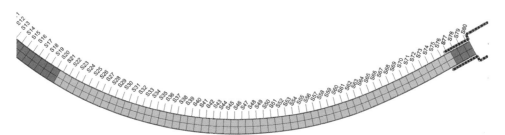

Figure 6.5. The tender solution. Plan

Segments S1 to S20 (where the tunnel is within the Miocene formations), with a total length of 96m, are slightly reinforced. Segments S21 to S77, which belong to the section of the tunnel with the most deformations and damage, with a total length of 273.6m (57 segments) are strongly reinforced and are separated from the outer tube by a waterproof elastic membrane, so that the two tubes behave independently.

The cross-section of the new tube consists of a slab with thickness ranging from 0.40m on the corners to 0.80m at the middle (0.60m average thickness), built by demolishing the upper part of the bottom filling, and a circular vault with 0.35m average thickness, built using the TBM tunnel as exterior formwork (figure 6.6).

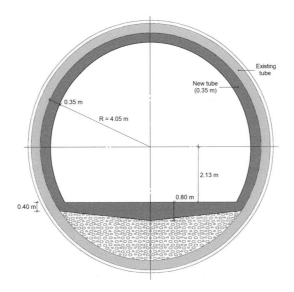

Figure 6.6.   The tender solution. Cross-section of the cast insitu segments

The temporary embankment installed along the tunnel alignment, is then partially removed, from level + 3.00 to level -3.50 (remaining about 6.5m), and installed a protection barrier on the riverbed, 35m wide, consisting of rockfill 0.80m thick over a layer of tout-venant 0.25m thick, in order to prevent the uplifting of the tunnel.

After the tender process, the works were awarded to the contractor Zagope SA.

6.5   ALTERNATIVE SOLUTION BASED ON THE TISB CONCEPT

After analysing the safety against earthquakes of the tender solution and having concluded that it was not adequate, the contractor developed an alternative solution based on the TISB concept (see, Chapter 5), with the supporting seismic studies carried out at the research center ICIST, of the University of Lisbon (Sousa-Oliveira et al. 2006). The author collaborated with the contractor as a consultant in the development of this solution (Pompeu-Santos 2005).

In the alternative solution, both the static behaviour and the seismic behaviour of the rehabilitated tunnel were explicitly considered. As for the static behaviour, in order to avoid the risk of uplifting the tunnel due to the reduction of stabilizing actions, the possibility of removing all the existing terrain over the tunnel was admitted.

On the alternative solution the main works planned are (figures 6.7 and 6.8):

a) Execution of a continuous tube, in reinforced concrete dully confined, inside the existing TBM tunnel, in an extension of 273.6m (the same as the most reinforced section in the tender solution);

b) Execution of three intermediate supports (S1 to S3), creating four spans 63.0 meters long each; the connections of the ends of the tunnel section, to a large block on the stiff ground (Miocene) on the west side and to the Metro station on the east side, function as rigid supports; the remaining stretch on the west side, 21.60m long, is within the stiff ground.

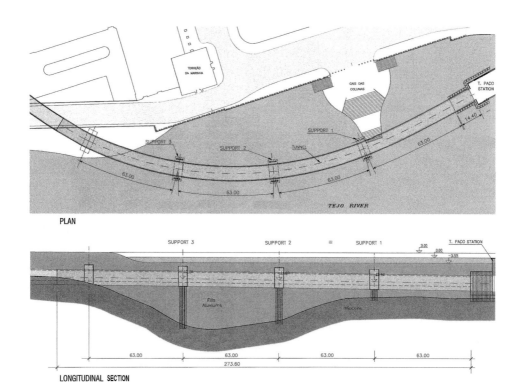

Figure 6.7.   Alternative solution. Plan and longitudinal section

The cross section of the new tube is similar to that of the tender solution, consisting of a circular vault with an average thickness of 0.35m and a slab with a thickness varying from 0.40m in the corners to 0.60m in the middle (0.50m average thickness) (see, figure 6.12), also built by demolishing the upper part of the existing concrete filling.

The intermediate supports are formed by groups of height reinforced concrete piles, anchored on the stiff soil, with 0.80m in diameter (two groups of four piles on each side), which are connected to the tunnel by anchoring them in large blocks that surround its upper and lateral sides (figure 6.8). The free height of the piles is about 7 meters on support S1, 22 meters on support S2 and 25 meters on support S3.

The blocks of the intermediate supports are 15m x 6m in plan and 13m high, and are formed by jet-grouting columns. Given the difficulty of executing jet-grouting columns under the tunnel, radial anchors complete the connection, anchoring the new tube to the sections of the piles inside the blocks (figure 6.8).

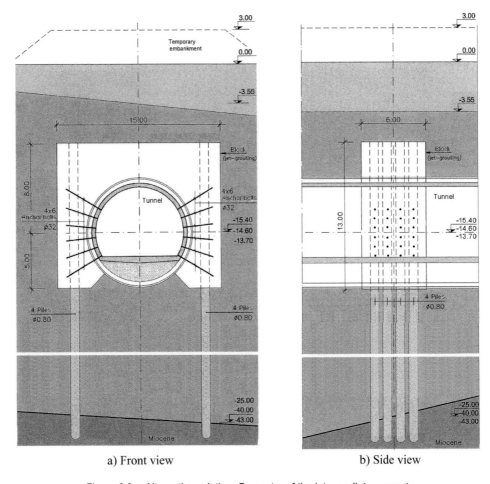

a) Front view   b) Side view

Figure 6.8.   Alternative solution. Geometry of the intermediate supports

The end block is also 15m x 6m in plan and is about 11m high, being also formed of jet-grouting columns (figure 6.9).

As the new structure is able to resist uplifting without taking into account the weight of the upper soil, the layer protection of the riverbed is not required.

However, as the contract did not explicitly provide for the submission of variants, the alternative solution was not accepted by the owner, who chose to implement the tender solution. Subsequently it carried out an additional contract for the seismic strengthening of the tunnel, developed by another designer, which consisted of curtains of reinforced concrete piles of large diameter, placed on both sides of the tunnel, anchored in the stiff soil.

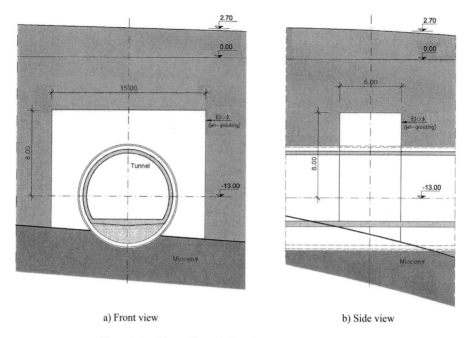

a) Front view   b) Side view

Figure 6.9.   Alternative solution. Geometry of the end block

However, the cost of the complementary contract was much higher than the increase in the cost of the alternative solution compared to the tender solution.

6.6   STRUCTURAL VERIFICATIONS

6.6.1   Static behaviour

Regarding the alternative solution, the respective reinforcements were calculated using linear models assuming simplified assumptions.

The new tube is capable of resisting, functioning as a longitudinal beam on five supports, the acting actions, without taking into account the resistance of the underneath soil (figure 6.10). The contribution of the existing tunnel for the strength of the structure is also neglected.

Thus, the new tube is able to withstand its self-weight, the weight of the existing tube and the weight of the soil above, taking into account the water pressure. It will also be able to prevent the tunnel from uplifting in the case the soil above is completely removed.

The cross section of the new tube will also be able to resist water pressure and horizontal reactions of the soil (figure 6.11).

Figure 6.12 shows the geometry and reinforcements of the new tube of the alternative solution.

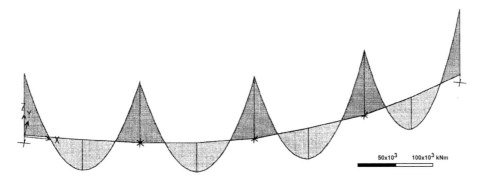

Figure 6.10. Alternative solution. Bending moments along the tunnel axis

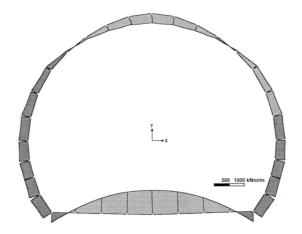

Figure 6.11. Alternative solution. Bending moments on the transverse section of the new tube

6.6.2 Seismic behaviour

The seismic behaviour of the tender solution and of the alternative solution were verified according to the Portuguese code on actions (INCM 1983), complemented by Eurocode 8-Part 1 (CEN 2004) and Eurocode 8- Part 5 (CEN 2002), and compared.

Based on the geologic-geotechnical conditions of the soil (figure 6.4), a soil profile was established for the area where the alluvia deposits are thicker. The main parameters on the dynamic response of the soil were calculated. Taking into account the results obtained, phenomena of liquefaction and cyclic mobility, locally or in a generalized way, were admitted.

As the mass of the surrounding soil is much greater than the mass of the structure, the dynamic behaviour of the soil-structure system will depend, essentially, on the properties of the soil. Therefore, the effects of an earthquake on the structure were treated as horizontal displacements imposed by the soil to the structure. The contribution of the existing TBM tunnel for the seismic behaviour of the structure was neglected.

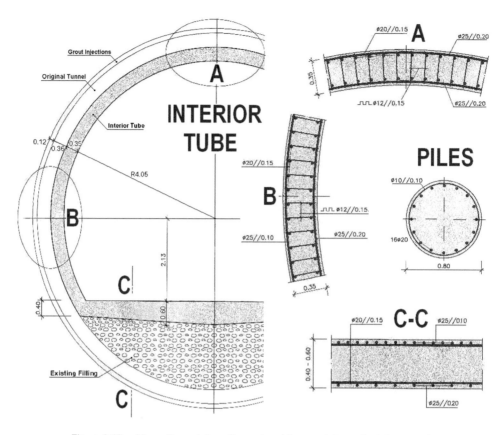

Figure 6.12.   Alternative solution. Geometry of the new tube and reinforcements

Three modes of deformation were considered: curvature deformations, cross-section ovalization and axial deformations.

Still according to the results obtained, a value of 0.20m was admitted for the maximum horizontal displacement at half height of the tunnel, in the thickest zone of the alluvia (between supports S2 and S3), with about 30m in height, that is, a distortion, d/h = 150, with linear variation up to the ends of the tunnel.

In the case of the tender solution, the horizontal displacements imposed by an earthquake, will translate into rigid body movements of the segments that form it.

Those movements lead to the opening or the closure of the joints between segments (depending on whether the movement is directed towards the outside or towards the inside of the curve, respectively), if the deformed is mainly of the type of flexion, or will rise to discontinuities between the joints, if the deformed is mainly of the shear type.

As the joints are 20 mm wide, in principle they can absorb distortions with an identical value.

Assuming that the imposed horizontal displacement of 0.20m will be absorbed in about 60m of tunnel, in the case the deformations are mainly of the shear type, discontinuities of the order of 15 mm will be obtained if the translation is uniformly distributed across the joints, but values much greater than 20 mm will be obtained if there is a concentration of deformations in only a few joints. In the case of bending-type deformations the situation will be similar, if the deformation is concentrated in only a few joints.

This level of translations will affect the integrity of the joints and, therefore, their tightness, as well as the alignment of the rails and, therefore, the safety of the trains inside the tunnel.

In the case of the alternative solution, the structure will behave as a horizontal continuous beam on supports (the great flexibility of the piles does not reduce horizontal displacements), being necessary to control the level of strains in the concrete of the new tube, taking into account the strains already imposed by the permanent loads.

Using a simplified linear model (figure 6.13), the maximum value obtained for the concrete strains in the new tube due to axial and bending deformations is $\varepsilon = 0.0018$.

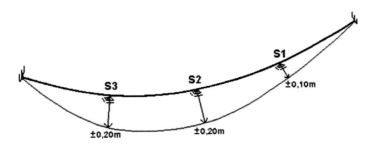

Figure 6.13. Model for checking the horizontal deformation of the tunnel in the alternative solution

When the movement of the soil is towards the inside of the curve, these strains are compression strains, and this value is perfectly acceptable, even for unconfined concrete, where $\varepsilon_{max} = 0.0035$. When the movement of the soil is towards the outside of the curve, the strains are tension strains, which only induces a slight instant cracking of the tunnel.

On both solutions the cross section of the new tube was also checked to the effect of the differential displacements between the top and the bottom (cross section ovalization).

Linear analysis was also used, considering a differential displacement of 0.05m to a height of 7m of soil, that is, 1/150 distortion (identical to that considered on the longitudinal analysis of the tunnel) (figure 6.14).

Using identical methodology, the maximum value obtained for the concrete compression strains is $\varepsilon = 0.0050$ in both solutions, which will occur in the vault/slab connection.

On the alternative solution this value is perfectly acceptable because, for the confining reinforcement used ($\Phi12//0.15$) $\varepsilon_{max} = 0.0100$, while in the tender solution that value is not acceptable, as concrete is not confined.

The piles of the intermediate supports of the alternative solution were also checked for the horizontal movements of the soil, being analysed as linear elements with the rotations restricted at the ends (the connections to the blocks at the top and to the stiff soil at the base).

The relative horizontal displacements between the ends of the piles were the same applied to the tunnel in the transverse direction. The maximum values for the strains are $\varepsilon = 0.0012$ for the longer ones (support S3) and $\varepsilon = 0.0096$ for the shorter ones (support S1), which are

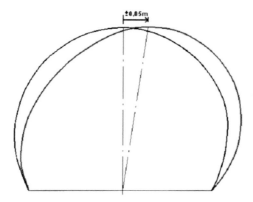

Figure 6.14.   Model for checking the cross-section of the new tube at differential displacements between the top and the bottom in both solutions

perfectly acceptable because, for the confining reinforcement used in the piles ($\Phi 10//0.10$), $\varepsilon_{\max} = 0.0120$.

The effects of soil movements in the direction of the tunnel axis (axial deformations) were also analysed. While in the alternative solution they have no influence, as the tunnel has longitudinal continuity and axial stiffness, in the tender solution, there may be movements of traction at the joints, with the risk of opening the joints between elements, aggravating the effects of the movements transverse to the tunnel.

Although the available information does not allow to evaluate if the phenomena of liquefaction and cyclic mobility of the surrounding soil will occur locally or in a generalized way, in the tender solution, localized situations (in 5 to 10 meters extension, for example), may cause the loss of support of entire elements, leading to large transverse displacements between adjacent elements, risking the tightness and integrity of the tunnel.

On the contrary, the alternative solution will be insensitive to these phenomena, since the structure would resist the localized or generalized loss of support, functioning as a continuous beam.

6.7 CONCLUSIONS

The section of the TBM tunnel of the Metro line between Baixa-Chiado and Praça do Comércio stations, in Lisbon, Portugal, built on soft ground in a seismic area, suffered an incident at the final stage of its construction, which has obliged to its rehabilitation.

At the initial stage of the rehabilitation process, the contractor questioned the reliability of the rehabilitation solution that has put to tender by the owner, namely its seismic behaviour, and developed an alternative solution based on the TISB concept, being the two analyzed and compared.

It has been concluded that the tender solution, in addition to not allowing the terrain over the tunnel to be removed, the safety against earthquakes present several weaknesses and it is not satisfactory, obliging to the adoption of additional strengthening measures.

In contrast, the alternative solution is able to withstand the weight of the tunnel, without taking into account the resistance of the ground below and to resist the uplifting of the tunnel in case the upper terrain was removed. In addition, the solution present adequate safety against earthquakes, also resisting in case of liquefaction or cyclic mobility of sand layers during an earthquake.

Thus, although the alternative solution is slightly more expensive than the tender solution, the cost-benefit ratio is much more favorable, as it avoids the need for any additional seismic strengthening measures, which globally lead to a much more expensive intervention.

It can therefore be concluded that the TISB concept provides very reliable and cost-effective solutions for the construction and rehabilitation of TBM tunnels on soft ground in seismic areas.

REFERENCES

CEN (2002), *ENV1998-5: Eurocode 8- Design Provisions for Earthquake Resistance of Structures. Part 5: Foundations, Retaining Structures and Geotechnical Aspects*. CEN, 2002

CEN (2004), *EN1998-1: Eurocode 8- Design of Structures for Earthquake Resistance. Part 1: General Rules, Seismic Actions and Rules for Buildings*. CEN, December 2004

Ferconsult (2002), *Prolongamento da Linha Azul. Poço da Marinha - Estação do Terreiro do Paço (Prolongation of the Linha Azul. Poço da Marinha - Terreiro do Paço Station)*. Ferconsult, Report (in Portuguese), July 2002

INCM (1983), *RSAEP: Regulamento de Segurança e Ações para Estruturas de Edifícios e Pontes (Safety Regulations and Actions for Structures of Buildings and Bridges)*. INCM, Decreto-Lei n° 235/83 (in Portuguese), 1983

INPI (2008), *Túnel de Comportamento Sísmico Melhorado (Tunnel of Improved Seismic Behaviour)*. INPI, Patente Nacional n° 103421 (in Portuguese), July 2008

Maranha-Neves E (2003), *Underground Works in Baixa de Lisboa*. Challenges of Engineering in Portugal, Odebrecht / Bento Pedroso, October 2003

Odebrecht (2003), *Lisbon Metro*. Challenges of Engineering in Portugal, Odebrecht / Bento Pedroso, Lisboa, 2003

Pompeu-Santos S (2005), *Reforço do Túnel do Metro do Terreiro do Paço. Avaliação da Solução Posta a Concurso e Apresentação de uma Solução Melhorada (Rehabilitation of the Terreiro do Paço Metro Tunnel. Evaluation of the Tender Solution and Presentation of an Improved Solution)*. Zagope SA, Internal Report (in Portuguese), November 2005.

Pompeu-Santos S (2007), *Tunnels Supported on Bearing Blocks*. IABSE Symposium Improving Infrastructure Worldwide, Weimar, Germany, September 2007

Pompeu-Santos S (2008a), *Tunnels of Improved Seismic Behaviour*. FIB Symposium Taylor Made Concrete Structures. New Solutions for Our Society, Amsterdam, Netherlands, May 2008

Pompeu-Santos S (2008b), *Túneis de Comportamento Sísmico Melhorado. O Conceito TISB (Tunnels of Improved Seismic Behaviour. The TISB Concept)*. Conference BE2008 (in Portuguese), Guimarães, Portugal, November 2008

Pompeu-Santos S (2009), *Túneis de Comportamento Sísmico Melhorado. O Conceito TISB (Tunnels of Improved Seismic Behaviour. The TISB Concept).* Magazine Ingenium (in Portuguese), May-June 2009

Pompeu-Santos S (2011), *TBM Tunnels of Improved Seismic Behaviour. The TISB Concept.* ITA-AITES World Tunnel Congress (WTC2011) Underground Spaces in the service of a Sustainable Society, Helsinki, Finland, May 2011

Pompeu-Santos S (2012), *TBM Tunnels of Improved Seismic Behaviour. The TISB Concept.* 15th World Conference on Seismic Engineering (WCEE2012), Lisbon, Portugal, September 2012

Pistone R, Cunha A (2014), *Tunnelling in Portugal.* CPT (Portuguese Commission for Tunnelling), Lisbon, Portugal, 2014

Sousa-Oliveira C et al. (2006), *Reforço do Túnel do Metro do Terreiro do Paço. Estudos de Suporte para a Comparação do Comportamento Sísmico da Solução Posta a Concurso com o de uma Solução Alternativa (Rehabilitation of the Terreiro do Paço Metro Tunnel. Support Studies for the Comparison of the Seismic Behavior of the Tender Solution with that of an Alternative Solution).* ICIST EP, Report nº 3/06 (in Portuguese), January 2006

Chapter 7

TBM tunnels for high-speed railway lines (HSRL) based on the TMG concept

Abstract

This chapter presents the application of the TMG ("Tunnel Multi Gallery") concept in the conceptual design of tunnels for HSRL (High-speed railway lines) built with the TBM technique.

The current TBM tunnel solutions for HSRL are presented and discussed.

Proposals for TBM tunnel solutions for HSRL based on the TMG concept, for different speed levels, are presented, as well as an example of application of these solutions to a tunnel for the HSRL Lisbon-Porto, in Portugal.

The advantages of using these approaches are shown.

7.1 INTRODUCTION

High-Speed Railway Lines (HSRL) are usually considered to be the lines of rail nfrastructure networks for passenger trains at speeds of at least 250 km/h.

The first high-speed rail system was created in Japan, in 1964, the Tōkaidō Shinkansen (from Tokyo to Osaka), widely known as the "bullet train", operating at speeds of 250 km/h.

Subsequently, HSRL have spread around the world, especially in the most developed countries and also in some emerging markets. HSRL moved to speeds of 300 km/h and today commercial speeds of 350 km/h are common.

Starting in 2003, China quickly became the world leader with a network of over 30 thousand km of high-speed railway lines today, more than two thirds of the world total. In Europe, Spain, is the leader (and second in the world) with more than three thousand km of HSRL network (Wikipedia 2021).

Although passenger high-speed trains have lower maximum speeds than jet planes, they offer shorter total travel times than air travel for short distances. They typically connect city centre rail stations to each other, while air transport connects airports, usually further away from city centers.

HSRL is best suited for travel times of about one to four hours that is until 800–900 km distances. Compared to jet planes, for these travel times, the process of check-in and going through airport security, as well as getting to and from the airport, makes the total air travel time longer than with HSRL.

With this move, the number of passengers transported annually worldwide by HSRL has increased very rapidly, being now practically identical to that carried by airlines (around 4.5 thousand million in 2019), when it was only a quarter in 2000 (Wikipedia 2021).

DOI 10.1201/9781003144861-8

HSRL operates essentially on Standard (UIC) gauge tracks of continuously welded rails with low gradients and large radius curves. Due to the very demanding geometric and safety requirements of the HSRL, the lay-out of the tunnels plays a decisive role in the performance of the lines.

The TBM (Tunnel Boring Machine) technique is the most common today in the construction of tunnels for HSRL, due to economic as well as environmental issues, allowing the reduction of the construction cost and construction time, and the environmental impact.

7.2 TBM TUNNEL CONCEPTS IN HSRL

In the basic conventional solution of TBM tunnels for HSRL systems, a single tube is used, with a filling at the base and two tracks placed side by side (figure 7.1). For access to the outside of the tunnel, pedestrian galleries are usually arranged along the tunnel.

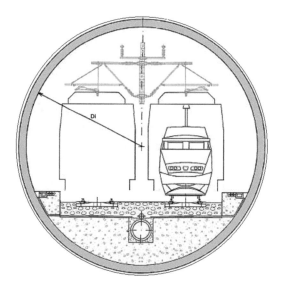

Figure 7.1. Cross-section of the basic conventional solution of TBM tunnels for HSRL

The inner diameter (Di) of the tunnel's cross-section basically depends on the maximum speed of the trains, the type of traffic (only passengers, or freight too) and the possibility of crossing the trains inside the tunnel. Tunnel cross-sectional areas between 82 m^2 and 112 m^2 have been adopted (González-Gutiérrez 2009).

In this solution, when necessary, the exchange of tracks by the trains is easily accomplished.

However, as mentioned, in order the safety requirements (EU 2008; UIC 2003) be met, the installation of the two traffic directions, side by side, in the same tube, is feasible only in short tunnels.

In long tunnels, due to the piston effect of the trains, limitations have to be imposed on the train speeds as well as the adoption of complex ventilation systems, in order to control the smoke flow inside the tunnel, in the event of a fire.

Thus, in the TBM tunnels in HSRL, the most common solutions are now the dual-tube single-track tunnels, in which the tunnel is formed by two tubes, one each for each traffic direction, interconnected by a system of cross-passages, regularly spaced, which provide access to emergency personnel and the passage of users from one tube to the other in an incident situation. As mentioned in Chapter 4, these are the cases, for example, of the Guadarrama Tunnel, in Spain, in the HSRL Madrid-Valladolid, 28.4 km long, and the Gothard-Base Tunnel, in Switzerland, 57.1 km long, the longest in the world.

In the event of an accident or fire inside one of the railway galleries, users will leave the incident train and walk to the other railway gallery through cross-passages, or through special galleries at "emergency stations", to be later rescued by another train.

Sometimes (in underwater tunnels, for example), the three-tube single-track tunnels are adopted, in which, in addition to the two railway tubes, a third tube (service tunnel) is adopted, placed between the other two and connected to them by regularly spaced cross-passages, which will be used to provide local access to emergency services and the evacuation of users to out of the tunnel. This is the case of the Channel Tunnel, between France and UK, one of the longest and complex tunnels in the world (see, Chapter 4).

In the event of an accident or fire inside the tunnel, users will leave the incident train and walk to the service tube through the cross-passages, where they will later be rescued by special wheeled vehicles.

In the dual-tube and the three-tube concepts, the cross-section of the railway tubes depend, primarily, on the maximum speed of the trains and on the type of traffic (passengers only, or freight also). For the traffic tubes, inner diameters of 7.5m to 8.8m are being adopted, while for the service tube an inner diameter of 4.8m has been adopted.

In both tunnel concepts, in very long tunnels, large caverns are also sometimes created, namely to house "emergency stations", for the stopping of damaged trains, and cross-overs, to allow trains to pass from one gallery to the other, in the event that one of the tracks becomes inoperative.

The need to build two or three tunnels and systems of cross-passages and of large caverns make these solutions very complex and expensive. In addition, despite the complexity of the safety systems used, as mentioned, the time required for rescue services to evacuate users in an incident situation can be very long.

Mono-tube dual-gallery tunnel solutions are also possible, as the Groenehart Tunnel, in the HSL-Zuid, in The Netherlands (see, also Chapter 4), in which a large diameter tunnel is used, with a filling installed at the base and a separating wall in the middle, in order to create two separate railway galleries placed next to each other.

Escape doors, regularly arranged in the separating wall of the tunnel, provide escape route for users in an incident situation occurred inside one of the railway galleries.

However, despite presenting several advantages compared to the solutions with two or three tubes, the conventional mono-tube dual-gallery tunnels also have some shortcomings.

As mentioned, in order to overcome these challenges, the TMG concept was recently developed by the author for railway tunnels, being particularly suitable for HSRL tunnels.

7.3   TBM TUNNELS FOR HSRL BASED ON THE TMG CONCEPT

7.3.1   General

As mentioned, the TMG ("Tunnel Multi Gallery") concept allows, with a suitable arrangement of the cross-section, a mono-tube TBM railway tunnel provides the capabilities there are usually offered by two or three parallel TBM tunnels and the adoption of a very reliable safety system for local access of the emergency services and evacuation of users in an incident situation, thus significantly reducing the construction cost and improving safety of the tunnel during operation.

The TMG concept allows in a mono-tube TBM tunnel with a cross-sectional diameter not too large, the installation of two independent and isolated railway tracks (each one for a direction of traffic) and a service (emergency) gallery below, which provides a dedicated passageway for the access of emergency personnel and the rescue of users in the event of an accident or fire inside the tunnel (Pompeu-Santos 2008, 2011, 2012, 2013, 2014a 2014b, 2016, 2018, 2019a, 2019b and 2020).

Based on the TMG concept, tunnels for HSRL are mono-tube TBM tunnels, inside which a slab and a separating wall are built, in order to create three galleries: two parallel and isolated railway galleries (ballastless), placed side by side, each for one track, and a service gallery below, being the three connected by vertical access galleries, regularly spaced along the tunnel. Both railway galleries are equipped with emergency walkways arranged on one side of the tracks, which give access to the vertical access galleries through escape doors.

Two alternative tunnel lay-outs can be adopted: Alternative A tunnels, based on Basic lay-out of the TMG concept, and Alternative B tunnels, based on Variant B of the TMG concept.

On Alternative A tunnels, vertical access galleries are arranged on the outer wall of the tunnel, on the exterior side of the railway galleries, with the emergency walkways being placed along the exterior side of the railway galleries.

On Alternative B tunnels, vertical access galleries are arranged in the middle of the tunnel, in the separating wall, with the emergency walkways being placed along the inner side of the railway galleries.

In both alternative tunnel solutions cross-overs can be arranged at appropriate distances along the tunnel, to allow trains to pass from one to the other railway gallery, in case of need. For that purpose, large openings are adopted on the separating wall of the tunnel, which will be closed by fire doors of the sliding type.

Also in both alternative tunnel solutions, in the areas of the portals, the separating wall is provided with openings, whose section decreases from outside to the inside of the tunnel, in order to reduce the pressure variation inside the trains, which allows a significant reduction of the cross-sectional area of the railway galleries (see, Chapter 5).

7.3.2   Lay-out of the tunnels

The inner diameter of the proposed tunnels (Di) will depend basically on the speed of the trains, being considered train speeds of 250 km/h, 300 km/h and 350 km/h (figures 7.2 to 7.4).

Although it is possible to consider tunnel solutions for higher train speeds, they are not viable due to infrastructure safety issues on conventional railway tracks, namely the risk of ballast projection or catenary instability at very high speeds (from 400 km/h).

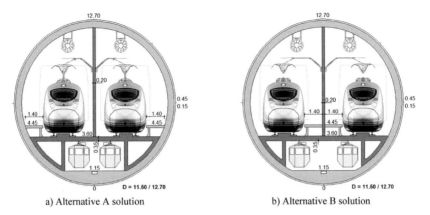

a) Alternative A solution   b) Alternative B solution

Figure 7.2.   Current tunnel cross-section for train speeds of 250 km/h

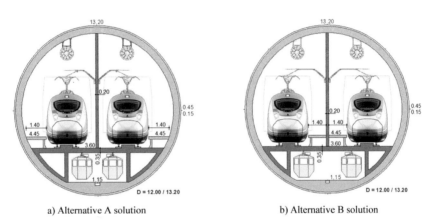

a) Alternative A solution   b) Alternative B solution

Figure 7.3.   Current tunnel cross-section for train speeds of 300 km/h

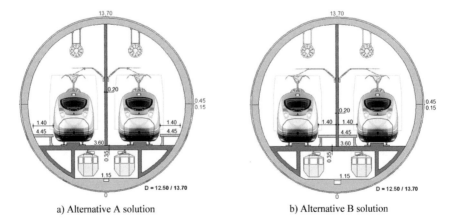

a) Alternative A solution   b) Alternative B solution

Figure 7.4.   Current tunnel cross-section for train speeds of 350 km/h

Taking into account the considerations made above, on Chapters 3 and 5, the inner diameter of the tunnels cross-section is 11.50m for speeds of 250 km/h, 12.00m for speeds of 300 km/h and 12.50m for speeds of 350 km/h.

Inside the TBM tunnel, an intermediate slab and a separating central wall, 0.35m and 0.20m thick, respectively, are built, creating two independent and isolated railway galleries, each for one track, and a service gallery below, 2.10m high.

The precast segments of the outer wall of the tunnels will, in principle, be 0.45m thick, that is, about 1/25 of the inner diameter of the tunnel. Thus, considering 0.15m clearance between the precast segments and the ground, to be injected with jet grouting, the excavated diameter of the tunnels will be 12.70m for train speeds of 250 km/h, 13.20m for train speeds of 300 km/h and 13.70m for train speeds of 350 km/h, which are common sizes for TBM tunnels today.

Table 7.1 summarizes the main dimensions of the cross-section of the tunnels.

Table 7.1. Main dimensions of the tunnels cross-section

	Train speeds		
	250 km/h	300 km/h	350 km/h
Inner diameter (m)	11.50	12.50	12.50
Excavated diameter (m)	12.70	13.20	13.70

In both alternative solutions, for any speed level, in the portal zones, the separating wall of the tunnel is provided with openings, the area of which decreases from the outside to the inside of the tunnel, and the edges are inclined by about 35° in relation to the horizontal plane, which allows to reduce the necessary cross-sectional area of the railway galleries by about 15 to 20% (see, Chapter 5).

The cross-sectional area of the railway galleries is therefore 38 m^2 for train speeds of 250 km/h, 44 m^2 for train speeds of 300 km/h and 50 m^2 for train speeds of 350 km/h.

As mentioned in Chapter 4, in the Groenehart Tunnel, in the HSL-Zuid, between Rotterdam and Amsterdam (in The Netherlands), a TBM mono-tube railway tunnel, traffic galleries with 50 m^2 cross-sectional area also were adopted, the separating wall having openings at the upper part, along the length of the tunnel, allowing train speeds of 350 km/h.

The slab is supported on the outer wall of the tunnel, on both sides, and on two longitudinal walls of small height. The slab is strengthened tranversely with ribs placed over the longitudinal walls, which also serve as anti-derailment devices. Thus, in the event of incidents of this type, the emergency walkways and the separating wall will not be affected.

In both alternative solutions the emergency walkways are, in principle, 1.40m wide, being made of precast concrete elements, placed next to each others.

The galleries under the emergency walkways are used as ducts for pipes and cables.

At regular intervals, on the order of 400 meters, escape doors are provided on the emergency walkways (placed on the outer wall or on the separating wall, in Alternative A and Alternative B tunnels, respectively), which give access to the service gallery below, through vertical access galleries, equipped with stairs (figure 7.5). As mentioned in Chapter 5, there is no significant gain in adopting less widely spaced escape doors.

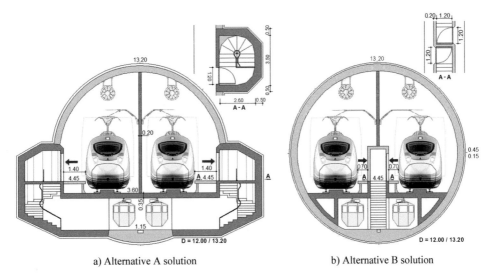

a) Alternative A solution   b) Alternative B solution

Figure 7.5.   Cross-section of the tunnels at vertical access galleries (speeds of 300 km/h)

Inclined platform lifts, disposed along those stairs, provide access for people with limited mobility.

Inside the service gallery, dedicated emergency vehicles of the EMEV type ("Emergency Monorail Electric Vehicles") are installed, which circulate suspended from the slab in two parallel lines (see, figures 7.2 to 7.5).

EMEVs will provide local access to emergency personnel (rescue personnel and fire fighters) and evacuation of users to out of the tunnel in the event of an accident or fire inside the tunnel. They are parked at the portal areas of the tunnel, in one or in both, depending on the length of the tunnel (see, Chapter 5, figure 5.26). In those areas, access buildings, provided with elevators and stairs, allow for access of users to the outside of the tunnel.

In those areas, access buildings, provided with elevators and stairs, allow for access of users to the outside of the tunnel.

Long tunnels (over 10 km in length) will be provided with cross-overs, to allow trains to pass from one railway gallery to the other, in case there is the need to deactivate a section of a track (maintenance works, broken train, etc.), which will be spaced 10 km at maximum.

Thus, in these areas, the separating wall will be provided with large openings (about 40m long and 6m high), protected with fireproof sliding doors, which will be opened to allow trains to move from one track to the other. In Alternative B tunnel solutions the emergency walkways will be interrupted in these areas, however they will have continuity through cross-walks at the level of the tracks.

In Alternative A tunnel solutions, the vertical access galleries will be built by locally disassembling the special precast lining segments mounted on these areas of the outer wall of the tunnel and casting new concrete walls insitu (see, Chapter 5).

As also mentioned in Chapter 5, in situations where there is water pressure around the tunnel, a kind of "cake" of waterproofing jet-grouting will be created on these areas, in order to allow the works to be performed safely.

In Alternative B tunnels, placing the vertical access galleries in the middle of the tunnel will cause a slight decrease (about 3 - 4%) of the cross-sectional area of the railway galleries in those areas and therefore a slight increase in the pressure variation inside the trains.

However, as the vertical access galleries are located outside the portal zones (the nearest ones are at a distance of about 400 m) their presence along the tunnel will not have significant influence on the comfort conditions inside the trains.

7.3.3   Technical systems

Drainage

The tunnels are equipped with a drainage system to drain the water that enters them (from the fire fighting system or from outside through the joints), dimensioned to the expected volumes.

Each railway gallery is provided with a line of sink drains, regularly distributed along the sides of the tracks, with the effluent being discharged into pipes placed along the underside of the slab, from which it is conducted for wells created along the bottom of the tunnel (see, figure 5.22). The effluent is then pumped out through pipes placed along one side of the service gallery.

The pavements of the traffic galleries have a gradient of 0.5% for the sink drain lines, with horizontal tubes crossing the longitudinal ribs at regular distances.

Fire fighting

For fire safety, the railway galleries of the tunnels are equipped with active detection and attack systems, acting jointly.

Heat sensors and smoke detection systems automatically activate an WFS ("Water Based Fire Fighting System") constituted by high pressure water mist nozzles, placed on both sides along the upper part of the railway galleries and grouped into sections of about 30m length (see, figure 5.24).

After this action, fire fighters, who access the fire site through the service gallery in EMEV ("Emergency Monorail Electric Vehicle") "trains", will easily extinguish the fire, using local fire hydrants.

Nozzles are distributed along the so-called "emergency stations", special places prepared in each railway gallery for the stopping of trains on fire, with a length at least equal to that of the longest trains that circulate in the tunnel and spaced 5 km at the most.

As soon as the burning train stops, the nozzle sections along the length of the carriages on fire will be automatically activated.

The fire hydrants are placed at regular intervals along the "emergency stations" on the outside of the emergency walkways.

Water pipelines placed inside the galleries under the emergency walkways supply pressurized water to the systems. As for the WFSs, the main lines supply secondary lines for each to each section and controlled by valves, which in turn feed the section's nozzles.

Ventilation

As the tunnels are equipped with WFSs ("Water Based Fire Fighting Systems") the ventilation system will be of the longitudinal type, with jet fans regularly distributed throughout the tunnel, which push the air and smoke in the direction of movement of trains in each railway gallery.

As in normal operation, the circulation of the trains is sufficient for the circulation of air inside the traffic galleries, the fans operate only in special situations: during maintenance operations or in case of fire (see, figures 7.2 to 7.5).

The service gallery is permanently pressurized through air injected by ventilators operating at both ends of the tunnel, in order to create an internal overpressure, that prevents the entrance of smoke in the vertical access galleries, when, in fire situations, the escape doors are opened for the escape of users.

7.3.4 Safety in operation

Procedures for safety in operation (safety concept) of the tunnels in HSRL are quite simple and very reliable.

In the event of activation a fire alarm, the train will continue to out of the tunnel, or at least until the next "emergency station", which will not be difficult to achieve, since they are only about 5 km apart. When the train stops inside the tunnel, users will leave the respective railway gallery by walking through the respective emergency walkway to the nearest escape door, from which they reach the service gallery below, downing the stairs inside the respective vertical access gallery. Inclined platform lifts running along the stairs provide access to people with limited mobility.

The service (emergency) gallery has two functions; (1) be a "safe place" for users in an incident situation, and (2) to be a dedicated passageway for the access of the emergency services (rescue personnel, fire fighters, etc.) to the incident site and the rescue of users, through emergency vehicles of the EMEV type. Thus, while fire fighters go up to respond to the situation, the rescue personnel organize the rescue of users to out of the tunnel in the same vehicles.

As mentioned in Chapter 5, EMEVs are monorail battery powered vehicles, remote controlled and equipped with smart cameras, which receive (and transmit) wireless signals from (and to) the Control Center of the tunnel, in principle with capacity for 20 people each.

EMEVs are grouped into "trains" of five units or more (depending on the maximum capacity of the passenger trains using the tunnel), being parked at the areas of the tunnel portals.

In each parking there are two EMEV "trains" available. In short and medium-size tunnels (less than 10 km long), EMEV "trains" are parked in only one of the portals, while in long tunnels (over than 10 km long), there are EMEV "trains" parked in both portals.

In very long tunnels (over 40 km long), in incident situations the four EMEV "trains" may park on the same portal, in order to reduce the rescue time for all users.

Long tunnels (over 10 km in length) will still be provided with cross-overs, to allow trains to pass from one track to the other, which will be spaced about 10 km.

Emergency personnel as well as users will never make use of the traffic galleries, therefore, in incident situations there will be no disturbance to the traffic flow in the non-incident gallery, which will continue to operate normally.

7.3.5   Costs

The construction costs of the tunnels can be estimated considering appropriate unit costs per cubic meter of excavation, according to the lay-out of the tunnel and the excavation conditions.

Taking into account that for TBM railway tunnels built in different parts of the world and under different conditions, unit construction costs between EUR 234 per cubic meter and EUR 328 per cubic meter were obtained, with an average value of EUR 287 per cubic meter (see, Chapter 2), unit costs of EUR 290 per cubic meter of excavation for Alternative A solutions and of EUR 280 per cubic meter of excavation for Alternative B solutions (lower value, as there is no need to build openings in the outer wall of the tunnel) can be adopted.

With excavation diameters of 12.70m in tunnels for train speeds of 250 km/h, 13.20m in tunnels for train speeds of 300 km/h and 13.70m in tunnels for train speeds of 350 km/h, which correspond to excavation cross-sectional areas of 127 m^2, 137 m^2 and 147 m^2 respectively, the construction costs will be, respectively, EUR 37, 40 and 43 million per km of tunnel in Alternative A solutions, and EUR 35, 38 and 41 million per km of tunnel in Alternative B solutions.

Thus, it can be considered that the construction costs of the tunnels are, on average, around EUR 40 million per km of tunnel. To these costs, the costs of the access ramps (when existing) will be added.

Taking into account the reductions obtained in the cross-sectional areas of the proposed tunnel solutions by at least 20% compared to those tunnels based on the TBM dual-tube concept, cost reductions of at least that amount will be achieved.

As for the costs of the access ramps, they will be similar to those of tunnels based on the dual-tube concept. In fact, although in mono-tube tunnels they are longer (in general, about 1.5 times longer) in dual-tube tunnels they are double, but about two thirds narrower each.

7.4   APPLICATION IN A RAILWAY TUNNEL AT THE HSRL LISBON-PORTO

The HSRL between Lisbon and Porto, in Portugal, is in the planning process. The best approach for drawing the line in the vicinity of Lisbon, over an extension of about 35 km, is to use the existing railway channel, improving two of the four existing tracks (Pompeu-Santos 2019a).

However, as between Alhandra and Vila Franca de Xira, in a distance of about 5 km, there are only 2 tracks and there is no space for the construction of the two additional tracks, a tunnel about 5 km long is necessary be built (figure 7.6).

The tunnel will be for mixed traffic (High-Speed passenger trains and freight trains). The maximum speed foreseen for trains inside the tunnel is 250 km/h.

The tunnel will run through Miocene formations along its entire length. It is relatively flat, with gradients of 1.0% in general and of 2.5% on the access ramps (figure 7.7).

The proposed solution is based on Alternative B tunnel solutions for HSRL at speeds of 250 km/h (figure 7.8).

The inner diameter of the tunnel will therefore be 11.50m and the excavated diameter will be 12.70m, using precast segments 0.45m thick and adopting a clear space 0.15m wide to be injected with jet-grouting.

Inside the tunnel, a slab placed at the bottom and a wall placed in middle of the tunnel is built, in order to create two independent and isolated railway galleries with 38 m^2 of cross-section, each for a direction of traffic, and a service gallery below, 2.10m high.

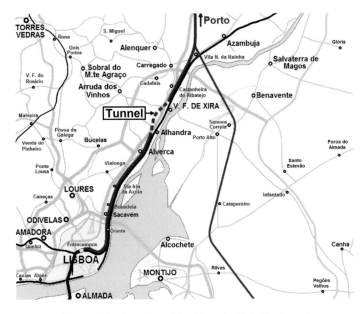

Figure 7.6.   Location of the Alhandra-V. F. Xira tunnel

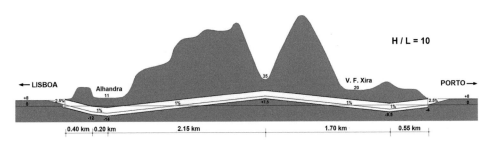

Figure 7.7.   Longitudinal section of the Alhandra – V. F. Xira tunnel

Vertical access galleries will be placed in the middle of the tunnel, on the separating wall, and the emergency walkways, 1.40m wide, will be placed on the inside of the railway galleries.

The tunnel will be provided with a drainage system in the railway galleries (see, figure 5.20b) as well a ventilation system of the longitudinal type, through jet fans (figure 7.8).

As the length of the tunnel is 5 km, there will be no "emergency stations" or cross-overs.

The traffic galleries will be provided with longitudinal ventilation through jet fans, while the service gallery will be provided with its own ventilation system, through ventilators installed at both ends.

On both emergency walkways, there are 12 escape doors of the rotating type, spaced 385m apart, which give access to the service gallery below, through vertical access galleries (figure 7.8b).

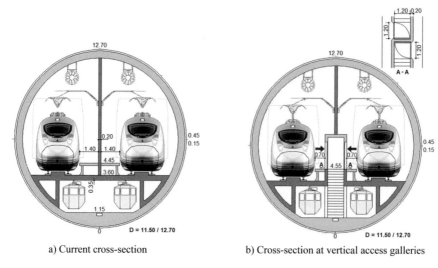

a) Current cross-section   b) Cross-section at vertical access galleries

Figure 7.8.   Cross-section of the tunnel

The service gallery will be equipped with two EMEV "trains" of five units each (see, figure 5.24), which are parked in one of the tunnel portals, to allow for access of the emergency personnel (rescue personnel and firefighters) to inside of the tunnel and the evacuation of users to out of the tunnel in the event of an accident or fire inside one of the traffic galleries.

In order to reduce the pressure variation inside the trains at the entrance or the exit of the tunnel, and thus to allow to reduce the cross-sectional areas of the railway galleries, in the portal zones the separating wall will be provided with openings, whose area decreases from outside to inside, and the edges will be inclined 35° in relation to the horizontal plane (figure 7.9) (see, Chapter 5).

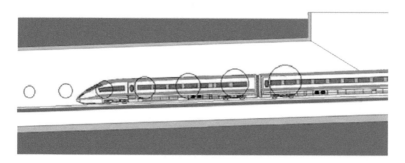

Figure 7.9.   The separating wall at the portal zones of the tunnel

As regard to the construction cost, according to the considerations made in section 7.4, as the tunnel length is 5 km, its cost will be about EUR 200 million, excluding the cost of the access ramps.

7.5 CONCLUSIONS

The TMG concept allows for obtaining very cost-effective TBM tunnel solutions for HSRL (High-Speed Railway Lines), that is, railway lines for passenger trains at speeds of 250 km/h or higher. In addition to the intrinsic advantages of TBM tunnels, the TMG concept provides significant cost savings and improvement of safety during operation.

Inside a mono-tube TBM tunnel of common diameter, a slab is built slightly above the bottom and a wall in the middle of the tunnel, in order to create two independent and isolated railway galleries and a service (emergency) gallery bellow, the three being connected by vertical access galleries. The railway galleries are provided with emergency walkways to allow users of an incident train to escape to the service gallery.

The separating wall in the middle of the tunnel allows the creation of pressure relieves in the portal zones, reducing the pressure variation inside the trains, thus allowing a significant reduction of the cross-sectional area of the railway galleries and thus of the diameter of the tunnel.

Two alternative tunnel lay-outs for the tunnels can be adopted: Alternative A solutions, based on the Basic lay-out of the TMG concept, in which the vertical access galleries are arranged on the outer wall of the tunnel, and Alternative B solutions, based on Variant B of the TMG concept, in which the vertical access galleries are arranged on the middle of the tunnel, at the separating wall.

The geometry of the tunnels cross-section will depend basically on the speed of the trains, thus, for train speeds of 250 km/h, 300 km/h and 350 km/h, the inner diameter will be 11.50 m, 12.00 m and 12.50 m, respectively, and the excavated diameter will (in principle) be 12.70 m, 13.20 m and 13.70 m, respectively.

For fire fighting, tunnels will be equipped with efficient WFSs ("Water Based Fixed Fire Fighting Systems"), constituted by high-pressure water mist nozzles, installed in "emergency stations", arranged along the tunnel, with spacing maximum of 5 km.

The ventilation systems of the railway galleries are very simple, of the longitudinal type, based on jet fans, which work only during maintenance operations or in incident situations.

With regard to the safety concept (safety in operation), an innovative concept is adopted, which is on the one hand quite simple and on the other hand very reliable. The service gallery at the bottom of the tunnel offers (1) a "safe place" for users in the event of an incident situation and (2) a dedicated passageway for access of the emergency personnel and rescue of users to out of the tunnel.

Inside the service gallery, remote controlled emergency vehicles of the EMEV type ("Emergency Monorail Electric vehicle"), provide access for the emergency personnel and the rescue of users. They are grouped in "trains" of five units or more (each EMEV with capacity for 20 people, in principle), being parked in one or both tunnel portals (two EMEV "trains" in one portal.

Long tunnels (over 10 km in length) will be provided with cross-overs, to allow trains to pass from one railway gallery to the other, which will be spaced 10 km at maximum.

With regard to construction costs, taking into account the average unit construction costs of the TBM railway tunnels and the cross-sectional areas of the tunnels, construction costs in the order of EUR 40 million per km of tunnel (excluding the cost of access ramps, when existing) can, in general, be considered, which represents a reduction of about 20% compared to the costs of conventional TBM railway tunnels.

It is, therefore, concluded that the TMG concept allows to obtain very cost-effective solutions of TBM tunnels in HSRL (High-speed railway lines), significantly reducing construction costs while greatly improving safety in operation.

REFERENCES

Bogaert P (2009), *Current and End Section Design Requirements for High-Speed Railway Tunnels*. Workshop Tunnels for High Speed Railway, Porto, Portugal, October 2009
Bopp R, Hagenah B (2009), *Aerodynamics, Ventilation and Tunnel Safety for High Speed Rail Tunnels*. Workshop Tunnels for High Speed Railway, Porto, Portugal, October 2009
EU (2008), *Safety in Railway Tunnel*. EU Decision 2008/163CE, January 2008
González-Gutiérrez A (2009), *Tunnels in the Cordoba-Malaga High Speed Line*. Workshop Tunnels for High Speed Railways, Porto, Portugal, October 2009
INPI (2009), *Túnel de Galeria Múltipla (Tunnel Multi-Gallery)*. INPI (in Portuguese), Portuguese Patent n. 103769, January 2009
Martin A, Trocellier G (2009), *Integration of Functionality Technology and Construction for HSL Tunnels*. Workshop Tunnels for High Speed Railways, Porto, Portugal, October 2009
Pompeu-Santos S (2008), *TISB and TMG Concepts- Application on a Proposal for the Railway Tunnel through the Gibraltar Strait*. IABSE Symposium Improving Infrastructure worldwide, Chicago, USA, September 2008
Pompeu-Santos S (2011), *Tunnels for Large Crossings: Challenges and Innovations*. IABSE/IASS Symposium Taller, Longer, Lighter, London, UK, September 2011
Pompeu-Santos S (2012), *Tunnels for High-Speed Railway Lines Based on TMG Concept*. IABSE Congress Innovative Infrastructures – Toward Human Urbanism, Seoul, Korea, September 2012
Pompeu-Santos, S (2013), *The TMG and TMF Concepts: the Right Approach for Large Tunnel Crossings*. 13th World Tunnel Congress (WTC2013): Underground, the Way to the Future, Geneva, Switzerland, 2013
Pompeu-Santos S (2014a), *High-Speed Railway Tunnels Based on TMG Concept*. 14th World Tunnel Congress (WTC2014) Tunnels for a Better Life, Foz do Iguassu, Brazil, May 2014
Pompeu-Santos S (2014b), *New Trends in the Tunnels of the Future*. IABSE Symposium Engineering for Progress Nature and People, Madrid, Spain, September 2014
Pompeu-Santos S (2016), *Sustainable TBM Tunnels for Tomorrow*. Second International Conference on Concrete Sustainability (ICCS2016), Madrid, Spain, June 2016
Pompeu-Santos S (2018), *The Long Traffic Tunnels of Tomorrow*. IABSE Symposium Tomorrow's Megastructures, Nantes, France, September 2018
Pompeu-Santos, S (2019a), *Grandes Projetos de Obras Públicas. Desafios Portugal 2030 (Major Public Works Projects. Challenges Portugal 2030)*. Scribe (in Portuguese), Lisbon, March 2019
Pompeu-Santos S (2019b), *The Long Rail and Road Tunnels of Tomorrow*. FIB Symposium Concrete- Innovations in Materials, Design and Structures, Krakow, Poland, May 2019
Pompeu-Santos, S (2020), *Innovative Concepts in TBM Tunnels*. Tunnel Engineering – Selected Topics (Chapter 7), IntechOpen, UK, April 2020
UIC (2003), *UIC Code 779-9: Safety in Railway Tunnels*. UIC, 2003
Wikipedia (2021), *High-Speed Rail*. Wikipedia, 2021

Chapter 8

TBM tunnel solution for a road crossing on soft ground in a seismic area based on the TISB and TMF concepts

Abstract

This chapter presents the application of the TISB and TMF concepts in obtaining an improved TBM tunnel solution for the construction of a road tunnel on soft ground in a seismic area, crossing the Tagus River, in Lisbon, Portugal.

After discussing different approaches, from a suspension bridge to an immersed tunnel and a conventional TBM tunnel, a detailed description of the proposed solution is presented.

This solution, in addition to providing improved structural safety in the event of earthquakes and adopting an innovative and very reliable system for the access of emergency services and the evacuation of users, its cost will be significantly lower than that any of the mentioned alternatives.

8.1 INTRODUCTION

The crossing of the Tagus River in Lisbon, Portugal, is currently carried out through two bridges: the 25 de Abril Bridge (initially called Salazar Bridge), open to traffic in 1966 (MOP 1966) and upgraded in 1998 (REFER 1999), which connects the southwestern area of Lisbon to Almada, in the south; and the Vasco da Gama Bridge, inaugurated in 1998 (GATTEL 1999), which connects the northeastern area of Lisbon to Montijo, in the east (figures 8.1 and 8.2).

The 25 de Abril Bridge is for mixed traffic, with two superimposed decks, a road deck on the upper floor and a rail deck on the lower floor. However, the road deck has been operating well above its capacity for several years, being a constraint for crossing the Tagus River in the western part of the city of Lisbon, where activities are more intense.

With 3 lanes in each direction, 3.25m wide each, the 25 de Abril Bridge has a flow capacity of around 100 thousand vehicles per day, in terms of AADT (Annual Average Daily Traffic).

However, with an average daily traffic of close to 150 thousand vehicles per day (Pompeu-Santos 2019c), the service provided by the bridge has been of poor quality for several years, forcing large congestion, particularly in the peak periods during the weekdays, in the morning in the south-north direction and in the late afternoon in the north-south direction.

As it is not possible to install more lanes on the bridge, national road authorities recognize the need to build a new road crossing in this area, with a view to decongesting the bridge.

To satisfy current needs (and probably also future ones), the new crossing must guarantee a flow capacity of at least 50 thousand vehicles per day, in terms of AADT.

DOI 10.1201/9781003144861-9

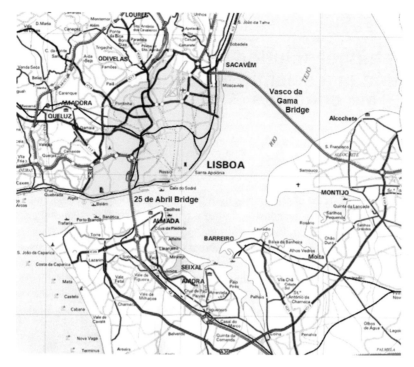

Figure 8.1.   The existing roadway network in the region of Lisbon, Portugal

a) 25 de Abril Bridge   b) Vasco da Gama Bridge

Figure 8.2.   The existing bridges crossing the Tagus River, in Lisbon, Portugal

Given the conditions of the existing road system on both banks of the Tagus River, it is foreseen that the new crossing is to be located west of the 25 de Abril Bridge and connect the CRIL on the north bank of the river to the A33 motorway (CRIPS), recently built on the south bank, allowing for the closing of the inner motorway ring of the region of Lisbon (ARIRL).

The geologic-geotechnical site conditions along the crossing are not yet known in detail. However, given the vicinity of the 25 de Abril Bridge, these conditions should not be significantly different from those of that bridge, which are well known.

Thus, the site is characterized by alluvial deposits of large thickness along the river bed, composed of various complexes of mud and sand, extending from elevation −20 until elevation −75. Underlying the alluvial deposits there are bed-rock formations composed of basalt and limestone, which extend through the north bank. On the south bank there are Miocene formations, composed mainly of sand and clay.

It is also to be noted the very prone-seismic conditions of the area; one must remember the 1755 Lisbon earthquake, one of the most destructive in history.

8.2 THE OPTIONS FOR THE ALGÉS-TRAFARIA CROSSING

8.2.1 Bridge versus tunnel

An Algés-Trafaria crossing can, in principle, be carried out through a bridge or a tunnel.

In the case of a bridge solution, it must be a cable supported bridge (suspended or cable-stayed type), with a large central span, such as the 25 de Abril Bridge. The bottom face of the deck should be very high, around elevation +60 (like the 25 de Abril Bridge) and the length will be over 3.5 km. Given the characteristics of the terrain, the foundations inside the river will very deep and complex.

The connections on the Lisbon side will be made on the west side of the Monsanto hill, while on the south bank they will be made over the Alto da Trafaria hills. In the Lisbon side they present several constraints, since it is a very urbanized area, in addition to the proximity to the Alto do Duque Fort (classified as a Monument of Public Interest). On the south bank, connections will also be very difficult and expensive, as it is in an area of high cliffs.

As for the visual impact, it will be overwhelming, conflicting with the image of the 25 de Abril Bridge. As for the cost, given its characteristics and the ground conditions, a suspension bridge will be a very expensive project, with a cost of more than EUR 1200 million.

In the case of a tunnel solution, it also presents significant difficulties, whatever the construction technique be used, due, on the one hand, to the difficult geologic-geotechnical conditions of the site, with thick alluvial layers and, on the other, because it is located in an area of great seismicity.

As for the connections at the ends, although they also present some difficulties, they are much simpler to make than on a bridge. As for the visual impact, it will, of course, be much less, not interfering with the current panorama of the Tagus River estuary. Regarding cost, it will be undoubtedly lower than that of a suspension bridge.

It can therefore be concluded that, in particular for reasons of environmental impact and cost, a tunnel is the preferred solution. Figure 8.3 shows the location of the proposed tunnel.

8.2.2 Immersed tunnel versus TBM tunnel

In the construction of the tunnel, in principle, two alternative techniques can be adopted: an immersed tunnel or a TBM tunnel.

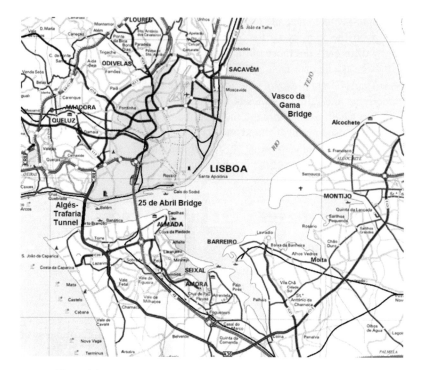

Figure 8.3.   Location of the Algés-Trafaria Tunnel in Lisbon, Portugal

As for an immersed tunnel, a feasibility study was carried out in 2001 (Cancio-Martins 2001). Figure 8.4 (Cancio-Martins 2001, adapted) shows the longitudinal section of the proposed immersed tunnel solution.

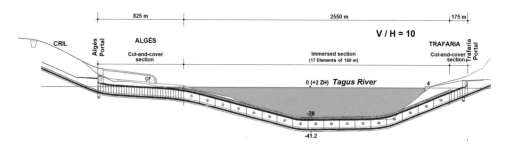

Figure 8.4.   Immersed tunnel solution for the Algés-Trafaria crossing. Longitudinal section

The tunnel is about 3.55 km long, being formed by a central immersed section with 2550 meters long and two cut-and-cover sections at the ends, with 825 meters and 175 meters long, on the north bank and the bank south, respectively, in addition to the access ramps, on both sides.

In terms of longitudinal profile, the immersed part of the tunnel is formed by three sections: a quasi-horizontal section, in the deepest zone (where the alluvial layers have greater thickness), running approximately between elevations −26 and −39, and two inclined sections, the one on the south side with 4.1% gradient and the one on the north side with 2.5% gradient. In the deepest zone, the alluvial layers under the tunnel are still 30 to 40 meters thick.

The cross section is that of a conventional immersed tunnel, formed by two road galleries 12.50 meters wide each and a service gallery arranged between them, 1.60 meters wide, so the cross section is approximately rectangular with 30.50 meters wide by 10.20 meters high (figure 8.5) (Cancio-Martins 2001, adapted).

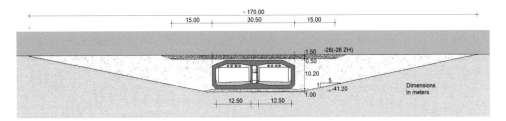

Figure 8.5. Immersed tunnel solution for the Algés-Trafaria crossing. Cross section

The road galleries house three lanes 3.5 meters wide each and edges 1.0 meters wide on both sides.

The immersed section is made up of 17 large precast elements, each 150 meters long, which are sunk in a trench previously dredged in the riverbed, and subsequently connected together through special rubber profiles that ensure tightness of the joints between precast elements. After injection of sand into the void space under the precast elements, the remaining space in the trench is filled with sand from the dredging, followed by the placement of a rock layer about 1.5 meters thick, for protection of the upper side of the tunnel.

The trench where the precast elements are sunk is about 170 meters wide at the top, since, as the sands have very little cohesion, the slopes have a gradient of 1 to 5 (20%). After dredging the bottom of the trench, it will be treated with vibro-compaction to reduce the risk of liquefaction of the sands on the riverbed in the event of earthquakes.

In the study it is recognized that the construction of the tunnel presents significant risks, associated, on the one hand, with the high probability of liquefaction of the sands, in the event of a strong earthquake, and, on the other hand, with the making the connections of the tunnel ends to the existing motorway network.

As for the connections at the ends, on the north bank, the tunnel passes under Avenida de Brasília, the railway line (Cascais Line), the Estrada Marginal and Praça D. Manuel II, also interfering with the cane of the Algés stream. However, there is no provision for connections to the local road network. On the south bank the tunnel interferes with the pier of Silopor, also crossing several secondary roads, as well as a residential area.

However, it is not mentioned how the tunnel will be connected to the CRIL. Now, given the level at CRIL and the overpass built in that area (which connects CRIL to Avenida de Brasília, south of the railway), this connection will be quite problematic.

In fact, as the tunnel access ramp and CRIL are not aligned and are almost parallel in height, the access ramp and the respective retaining walls have to be extended for several hundred meters (figure 8.3).

The estimated cost for the tunnel was EUR 480 million euros, that is, about EUR 650 million at current prices (update factor 1.36), although, as mentioned, important works are not included, so the cost of the tunnel will be significantly higher.

The immersed tunnel solution will also have some significant environmental impact, namely along the riverbed and in the connection areas at the ends.

As for a TBM tunnel solution, the environmental impact will be almost insignificant. However, although it has not been the subject of specific studies, it also presents some constraints.

So, although there should be no problems with the realization of the connections at the ends, either on one bank or the other, to avoid problems with the alluvial behavior in the event of earthquakes, the tunnel will, in principle, be bored through the rocky substrate, so it will have to go very deep, around elevation −100. As a result, it will be very long (about 7 km in length) to be inserted into the existing road network, both on the Lisbon side and on the south bank, so its cost will be very high.

However, the application of the TISB and TMF concepts (EPO 2011; INPI 2008, 2009) (see, Chapter 5) allows for a TBM tunnel to be bored through the alluvial layers and provided with adequate seismic safety, and at a cost significantly lower than that of a conventional TBM tunnel or that of an immersed tunnel. It will also be provided with an innovative and very reliable safety in operation concept (safety concept), for local access of emergency personnel and evacuation of users to out of the tunnel in the event of an accident or fire inside the tunnel.

Below is a proposal for the Algés-Trafaria crossing based on the above mentioned concepts (Pompeu-Santos, 2009, 2010a, 2010b, 2011a, 2011b, 2012, 2013a, 2013b, 2014, 2016, 2018, 2019a, 2019b, 2019c and 2020).

8.3   TBM TUNNEL SOLUTION BASED ON THE TISB AND TMF CONCEPTS

8.3.1   Plan and longitudinal section of the tunnel

The proposed Algés-Trafaria tunnel presents a north-south orientation, consisting in plan of two straight sections, connected by a large radius curve (figure 8.6). It will be 5.0 km long, of which 2.25 km under the river, 1.6 km on the north bank and 1.15 km on the south bank (figure 8.7).

In terms of longitudinal profile, the tunnel consists of five sections, with a maximum gradient of 4.8% (figure 8.7).

The tunnel runs through the alluvial deposits along most of its length under the river. Under the river the minimum cover of the tunnel will be identical to the excavation diameter, so, as the bottom of the river reaches elevation −28, in its deepest part, the tunnel bottom will be at elevation −59 (figure 8.7).

On the south bank, a section of motorway, about 3 km long, between the tunnel portal and the current end of the A33 highway, as well as the connection node with the IC20 motorway, will also be built (figure 8.8).

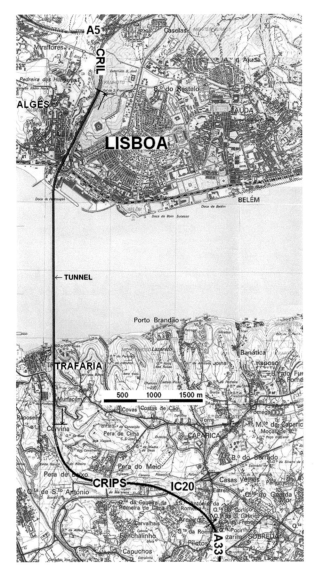

Figure 8.6.   Proposed TBM tunnel solution for the Algés-Trafaria crossing. Plan

On the north bank, in addition to the insertion into CRIL, the tunnel will be provided with connections to the local road network at the Algés low area, the Estrada Marginal (in both directions), as well as the Praça D. Manuel II (figure 8.9).

Due to the high water pressure under the river (about 6 bar), the tunnel will be bored by a TBM with front support (an EPB or a Mixshield, depending on a more refined characterization of the ground), starting on the South portal, where there is enough space for the assembling

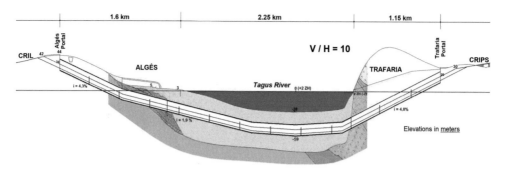

Figure 8.7.   Proposed TBM tunnel solution for the Algés-Trafaria crossing. Longitudinal section

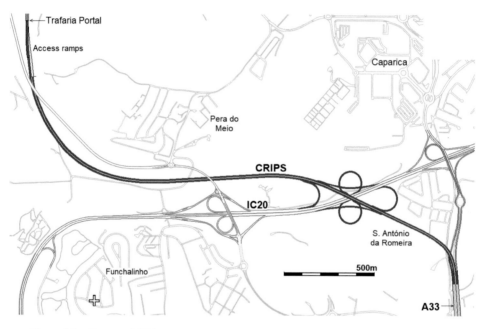

Figure 8.8.   Proposed TBM tunnel solution for the Algés-Trafaria crossing. South connections

of the machine. With the excavation rates currently offered by these types TBMs, the tunnel will be excavated in about one year.

The access ramps at the ends of the tunnel have a special lay-out to make the transition between the superimposed road galleries and the lanes placed side by side of the motorway (see, Chapter 5). As the alignment has curves at the ends of the tunnel, Alternative B solution will, in principle, be adopted (see, figure 5.13).

The connections to the local road network in the low area of Algés will be built using the top-down method of the cut-and-cover technique, using reinforced concrete secant piles, anchored on the stiff soil. In the deepest areas of these sections, intermediate slabs will be built, to act as lateral supports.

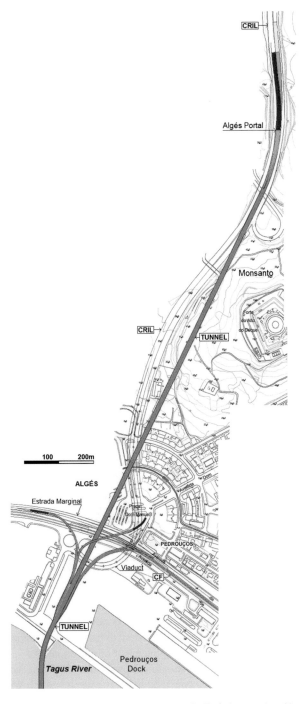

Figure 8.9. Proposed TBM tunnel solution for the Algés-Trafaria crossing. North connections

8.3.2 Cross-section of the tunnel

As mentioned, the proposed TBM tunnel solution will be built using the TBM technique, based on the TISB and TMF concepts.

The cross-section of the tunnel will therefore be circular, housing three superimposed galleries: two road galleries, each one with two lanes (three lanes, if necessary), and a service (emergency) gallery at the bottom.

The inner diameter of the TBM tunnel is 14.20m, so using precast segments 0.55m thick (about 1/25 of the inner diameter) and a gap of 0.15m thickness between the precast segments and the excavation surface, to be filled with jet-grouting injections, the diameter of the excavated tunnel will be 15.60m (figure 8.10a).

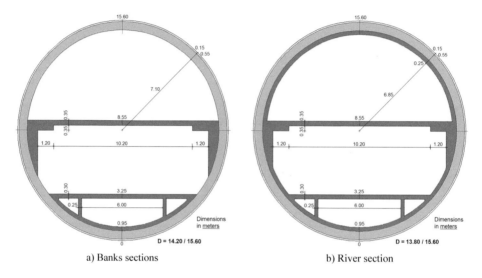

a) Banks sections   b) River section

Figure 8.10.   Cross-section of the tunnel

Although it is a large diameter, as mentioned in Chapter 4, TBM tunnels with much larger diameters have already been built, such as the SR99 Tunnel in Seattle (USA), built using an EPB 17.50m diameter, and the Northern part of the Tuen Mun-Chek Lap Tunnel in Hong Kong (China), built using a Mixshield 17.60m diameter. Studies for a Mixshield 19.25m diameter to be used in the Orlovski Tunnel, under the Neva River, in Saint Petersburg, Russia, have also been developed (Herrenknecht, 2011), although it was not built.

This is the tunnel solution adopted in the sections at the banks.

In the section inside the river, where the tunnel crosses alluvial layers, in a length of about 2.4 km (see, figure 8.7), the TISB concept is adopted, thus the TBM tunnel adopted in the banks is internally strengthened, with a view to improving its seismic behavior, ie, the improvement of resistance and ductility in the event of earthquakes.

Thus, inside the TBM tube, a cast in-situ concrete tube of 0.25m wall thickness is built (figure 8.10b), provided with dully confined reinforcements, which will be built using the outer tube as an external formwork and the aid of a mobile interior formwork. The in-situ concreted elements that are part of the current solution will be part of the inner tube.

Inside the tunnel, two reinforced concrete slabs are subsequently built, one placed approximately half height of the tunnel and the other placed just above the bottom of the tunnel, 0.35m and 0.30m thick, respectively (figures 8.10a and 8.10b).

The upper slab is thickened at the ends, in sections 1.20m long, so its thickness is 0.70m in these areas. It is supported laterally on continuous corbels connected to the precast segments of the TBM tube.

The corbels and the inner tube are connected to the TBM tube by means of reinforcement bars anchored on the intrados of the precast segments (see, Chapter 5).

The lower slab is supported on two longitudinal small cast in-situ concrete walls and on the tunnel wall at both ends.

Each road gallery houses two lanes of 3.50m wide each, an emergency lane 2.50m wide and an inner edge 0.70m wide, making a total road width of 10.20m, as well as emergency walkways on both sides, 1.20m wide each (figure 8.11a). The emergency walkways in the lower gallery are slightly elevated.

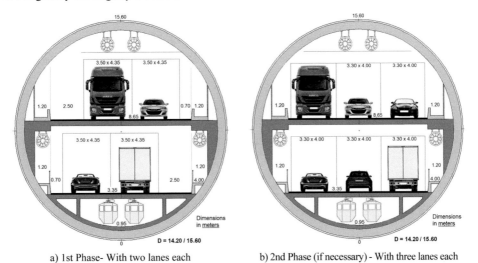

a) 1st Phase- With two lanes each   b) 2nd Phase (if necessary) - With three lanes each

Figure 8.11.   Lay-out of the traffic galleries of the tunnel

These geometric characteristics and a speed limit of 90 km/h give the tunnel a flow capacity of around 80 thousand vehicles per day, in terms of AADT.

The pavements of the traffic lanes consist of layers of bituminous with an average thickness of 0.10m, so that the total height of the road galleries is 4.85m.

By reserving a 0.50m clearance for the signaling installation, the maximum height for vehicle traffic will be 4.35m, so the tunnel has an efficiency index with an impressive value of 0.56, which is 1.33 times higher than the average efficiency indexes of the dual-tube and the mono-tube dual-deck road tunnels presented in Chapter 4, which is 0.42.

In a second phase (if traffic justifies it), although with some restrictions, the road galleries will be adapted for three lanes each, 3.30m wide and with a clear height of 4.00m each (figure 8.11b), and the speed of the vehicles limited to 70 km/h, which will increase the flow capacity of the tunnel to around 100 thousand vehicles per day, in terms of AADT.

The height of the service gallery is 2.00 meters, which, although relatively low, is sufficient, as the number of users will always be relatively small.

Each road gallery is provided with 12 escape doors of the rotating type, disposed on one side of the tunnel, which give access to vertical access galleries, 2.6m x 3.5m internal section, disposed parallel to the tunnel axis and spaced 385 meters apart (therefore less than the 500 meters allowed by EU legislation), and provided with stairs, for access to the service gallery bellow (figure 8.12).

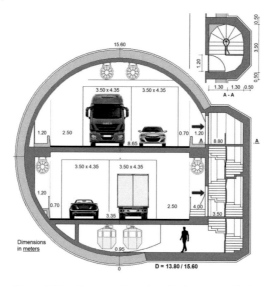

Figure 8.12.   Cross-section at vertical access galleries

Inclined platform lifts, arranged along the stairs, will provide access for people with limited mobility.

In the service gallery, at the base of each vertical access gallery, "safe rooms" with about 30 m^2 in plan, are available for users await to be rescued.

Vertical panels placed along the outside of the emergency walkways of the upper road gallery create longitudinal galleries for housing pipes and cables, and technical equipment.

As mentioned in Chapter 5, vertical access galleries will be built by disassembling special segments in steel, installed in those areas of the outer wall of the tunnel, and building (after the excavation of those galleries) reinforced concrete walls in-situ, which will be connected in continuity to the precast segments of the referred wall.

As also mentioned in Chapter 5, in the situations where there is water pressure around the tunnel, injections of jet-grouting will be carried out on those areas, creating a kind of "cake" of grout in the surrounding area, inside which the excavation will be carried out, ensuring work in safe conditions.

8.3.3 Structural safety

As for the structural safety of the tunnel, although the geotechnical characteristics of the ground are not well known yet, some considerations can be made.

On the banks sections, where the tunnel crosses stiff ground, having a wall thickness of 0.55m, it is expected that the tunnel will behaves in elastic regime, under both the static (self-weight, earth pressure, etc.) and the seismic actions. Under earthquakes, the tunnel will move together with the surrounding ground.

On the section within the river, where the tunnel crosses alluvial layers, the behaviour of the tunnel will be dependent on the characteristics of the surrounding soil along its length, as a result of the soil-structure interaction.

However, the inner tube provides a great increase in the resistance of the tunnel (both in the plane of the tunnel and in the tunnel cross-sections), on the one hand due to the great increase in the wall thickness (from 0.55m to 0.80m), and on the other hand due to the resistance provided by the internal reinforcements. Thus, the tunnel is expected to behave in elastic regime under the static actions.

Under earthquakes, the tunnel is expected to withstand large deformations in inelastic regime, which will be possible due to the increased deformation capacity of the concrete provided by the confinement reinforcement, which can reach 1.0% in bending.

Thus, the ductility provided by the confinement reinforcements of the inner tube, allowing high deformations in the confined concrete, makes the tunnel capable of withstanding large inelastic deformations for short periods of time, without significantly affecting its functionality and water tightness.

Figure 8.13 shows the reinforcement scheme of the upper slab and the connections to the precast segments.

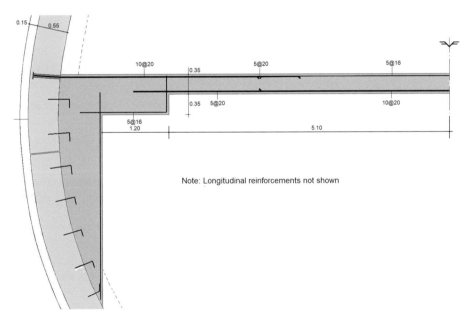

Figure 8.13. Lay-out of the upper slab reinforcement and connections to the precast segments

8.3.4 Technical systems

Drainage

As mentioned in Chapter 5, the traffic galleries are provided with a drainage system constituted by sink drains regularly distributed along the outside, the effluent being discharged into a tube placed along the underside of the slab, from which it is conducted to wells created along the bottom of the tunnel (figure 8.14).

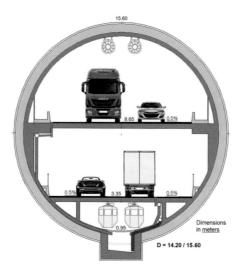

Figure 8.14.   Drainage system of the tunnel (at the river section)

The effluent collected in the wells is then pumped out to tunnel through pipes placed along the top of the service gallery.

If there are discharges of fuel or other contaminating liquids in the pavement of the traffic galleries, it will be possible to remove them from the wells.

Fire fighting system

As mentioned in Chapter 5, for fire fighting, both traffic galleries of the tunnel are equipped with fire fighting systems of the WFS type ("Water Based Fire Fighting System"), acting jointly with automatic detection systems.

Heat sensors and smoke detection systems automatically activate high pressure water mist nozzle systems, evenly distributed throughout the road galleries and grouped into sections 30 meters long (figure 8.15), drastically lowering the temperature at the fire site.

After this action, fire fighters, who come through the service gallery on EMEV "trains", will then be able to easily extinguish the fire.

Hydrants placed at regular intervals on both sides of the road galleries allow fire fighters to access the pressurized water network.

Fire pipelines disposed along the road galleries will provide pressurized water to the systems.

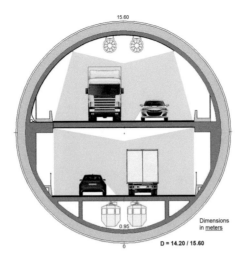

Figure 8.15. Fire fighting systems of the road galleries (at the river section)

Ventilation

As also mentioned in Chapter 5, the traffic galleries are equipped with longitudinal ventilation systems through jet fans, which push air and smoke in the direction of traffic, always ensuring fresh-fair from behind the vehicles. They act in two different situations: when the pollution particles are above the permitted levels or in fire situations.

The service gallery is permanently pressurized by means of ventilators installed at both ends, in order to prevent smoke from entering the vertical access galleries through the escape doors when, in fire situations, they are opened for the escape of users.

On the tunnel portals, at the exits of the traffic galleries, the polluted air is conducted to treatment plants. On figure 8.16 the exits from both road galleries are shown.

8.3.5 Safety in operation

For safety in operation (safety concept), the traffic galleries are provided with emergency walkways in both sides along its full length 1.20m wide. In the event of accident or fire inside one of the road galleries, users will leave this gallery by walking through the emergency walkways to the nearest escape door, from which they access the service (emergency) gallery below, downing the stairs of the respective vertical access gallery. Inclined platform lifts running along the stairs provide access to people with limited mobility.

Within the service gallery, EMEV "trains" (see, Chapter 5, figure 5.26), which are parked at the South portal of the tunnel and circulate suspended from the lower slab in two parallel lines, allow emergency services (rescue personnel and fire fighters) to access the site of the incident.

In the event of a fire, while the rescue team organizes the rescue of the users, fire fighters go up to the incident gallery to finish the extinguishment of the fire.

Longitudinal section   **Front view**

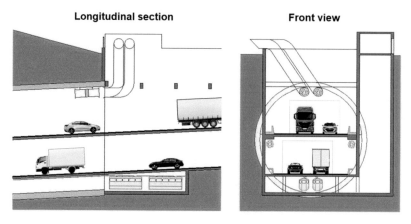

a) Exit of the upper road gallery (Trafaria portal)

Longitudinal section   **Front view**

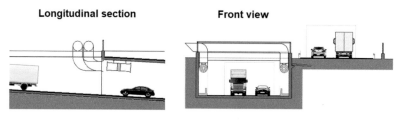

b) Exit of the lower road gallery (Algés portal)

Figure 8.16.   Ventilation systems at the portals of the tunnel

Users will be evacuated via the EMEVs to the Trafaria portal area, from where they access the outside of the tunnel through an access building equipped with elevators and stairs, located next to the Control Center of the tunnel (figure 8.16a).

As mentioned in Chapter 5, EMEVs are monorail unmanned vehicles, remote controlled, designed for a capacity of 20 people each, being grouped in two "trains" of two units each, thus, the global rescue capacity is of 80 people. The emergency personnel as well the tunnel users will never make use of the traffic galleries.

It is, therefore, a very reliable evacuation system, since the circulation of emergency vehicles will not depend on the traffic conditions inside the tunnel. In addition, the traffic flow in the non-incident gallery will not be affected for any incident inside the tunnel.

8.3.6 Estimated cost and financing

The cost of the TBM tunnel was estimated considering a unit tunnel cost per cubic meter of excavation, adopting a value appropriate to the characteristics of the tunnel and the conditions of the site.

Considering that in road tunnels built in different locations and conditions, unit costs between EUR 286 per cubic meter and EUR 545 per cubic meter of excavation were obtained, with an average value of EUR 406 per cubic meter (see, Chapter 2), it is appropriate to adopt for this tunnel a unit cost of EUR 420 per cubic meter of excavation.

With an excavation diameter of 15.60 meters, which corresponds to an excavated cross-section area of 191 square meters, the unit cost of the tunnel will be around EUR 80 million per km. As the length of the tunnel is 5.0 km, the cost of the TBM tunnel will be about EUR 400 million. This value is significantly lower than the cost of the immersed tunnel solution, which, as mentioned, will be, at least, about EUR 650 million (at current prices).

To the cost of the TBM tunnel, the costs of the access ramps at both ends will be added, as well as the costs of the connections to the existing network, which can be roughly estimated at EUR 50 million at each side (however, lower than those in the case of an immersed tunnel), so the total cost of the work will be in the order of EUR 500 million.

The construction of the tunnel may be the carried out by the concessionaire for the Tagus crossings in the Lisbon region, being financed through the tolls collected on the three crossings (the new tunnel and the two existing bridges), without charges to the State Budget or contribution from European funds. Considering a construction time of four years and an interest rate of 5 – 6%, with toll revenues at the level currently generated by the two existing crossings (around EUR 75 million annually), the new crossing will be paid in about 10 years of service.

8.4 CONCLUSIONS

It is commonly recognized the need to build a new road crossing on the Tagus River in Lisbon, to decongest the 25 de Abril Bridge, which has been operating well above its capacity for several years, and which should guarantee a capacity of at least 50 thousand vehicles per day, in terms of AADT.

The new crossing will connect the motorway A33 (CRIPS), at Monte da Caparica, on the south bank of the river, to the motorway A36 (CRIL), in Algés, on the north bank, to close the inner motorway ring of the Lisbon region (ARIRL). For environmental and cost reasons this crossing must be made through a tunnel.

The construction of the tunnel presents significant challenges, due to, on the one hand, the adverse geological and geotechnical conditions of the crossing site (thick alluvial layers), and, on the other, the strong seismicity of the area of Lisbon, where the work is being carried out.

A proposed TBM tunnel based on TISB and TMF concepts was developed. It will be 5.0 km long between portals. The cross section will be circular, with two superimposed road galleries and a service (emergency) gallery at the base. The inner diameter is 14.20 meters, resulting in an excavated diameter of 15.60 meters.

In the section inside the river, where the tunnel passes through layers of alluvium, it will be strengthened by means of a cast in in-situ tube, provided with dully confined reinforcements, in order to ensure the tunnel the necessary resistance and ductility in the event of earthquakes.

Each of the road galleries has two lanes, 3.50 meters wide and 4.85 meters high, an emergency lane 2.50 meters wide and an inner edge 0.70 meters wide, totaling a road width of 10.20 meters, plus emergency walkways on both sides 1.20 meters wide each, allowing a flow capacity of around 80 thousand vehicles per day, in terms of AADT.

In a second phase (if the traffic justifies it) the road galleries can be adapted to three lanes each, 3.30 meters wide and 4.00 meters high each, which will increase the flow capacity of the tunnel to around 100 thousand vehicles per day, in terms of AADT.

This lay-out of the tunnel's cross-section allows an efficiency index with and impressive value of 0.56, which is 1.33 times higher than the average efficiency indexes of noteworthy dual-tube and mono-tube dual-deck road tunnels, as shown in Chapter 4.

Both road galleries are equipped with escape doors of the rotating type, spaced 385 meters apart, which give access to the service gallery below, through vertical access galleries.

The service gallery is equipped with remotely controlled emergency vehicles of the EMEV type ("Emergency Monorail Electric Vehicle"), an innovative and very reliable safety concept for the access of emergency personnel to the inside the tunnel and evacuation of users, in the event of accident or fire inside the tunnel.

The EMEVs are grouped into two "trains" of two units each, which are parked in the South portal of the tunnel, next to the access building where users will have an exit to the outside of the tunnel, which is an integral part of the Control Center of the tunnel.

The estimated cost of the tunnel is EUR 400 million (excluding the cost of the access ramps and connections on both sides), well below the cost of a conventional TBM tunnel solution or an immersed tunnel.

It is, therefore, concluded that the TISB and TMF concepts allow the construction of TBM road tunnels on soft ground in seismic areas with adequate structural safety and provided with an innovative and very reliable safety in operation concept, in addition to be quite economical.

REFERENCES

Cancio-Martins (2001), *Nova Travessia do Tejo; Corredor Algés-Trafaria; Solução em Túnel Imerso (New Tagus Crossing; Algés-Trafaria Corridor; Immersed Tunnel Solution)*. J. L. Cancio Martins Lda, Report (in Portuguese), January 2001

EPO (2011), *Tunnel Multi-Storey*. EPO (European Patent Office), European Patent n. 2317074, November 2011

EU (2004), *Safety in Roadway Tunnels*. EU Directive 2004/54CE, April 2004

GATTEL (1999), *Ponte Vasco da Gama*. GATTEL (in Portuguese), 1999

Herrenknecht (2011), *Pioneering Tunnelling Technology for Underground Transport Systems*. Herrenknecht AG, 2011

INPI (2008), *Túnel de Comportamento Sísmico Melhorado (Tunnel of Improved Seismic Behaviour)*. INPI, National Patent n. 103421 (in Portuguese), July 2008

INPI (2009), *Túnel de Piso Múltiplo (Tunnel Multi Floor)*. INPI, National Patent n. 103748 (in Portuguese), August 2009

MOP (1966), *A Ponte Salazar (Salazar Bridge)*. MOP (in Portuguese), 1966

Pompeu-Santos, S (2009), *The TISB and the TMF Concepts: Application on a Proposal for a Roadway Tunnel Crossing the Tagus River in Lisbon*. IABSE Symposium Sustainable Infrastructure, Bangkok, Thailand, September 2009

Pompeu-Santos, S (2010a), *New Developments in Roadway Tunnels. The TISB and TMF Concepts*. 16th World Road Meeting (IRF2010), Lisbon, May 2010

Pompeu-Santos, S (2010b), *New Developments in Tunnel Building. The TISB, TMG and TMF Concepts.* IABSE Symposium Large Structures and Infrastructures for Environmentally Constrained and Urbanized Areas, Venice, Italy, September 2010

Pompeu-Santos, S (2011a), *TBM Tunnels of Improved Seismic Behaviour. The TISB Concept.* World Tunnel Congress (WTC2011) Underground Spaces in the Service of a Sustainable Society, Helsinki, Finland, June 2011

Pompeu-Santos, S (2011b), *Tunnels for Large Crossings: Challenges and Innovations* IABSE/IASS Symposium Taller, Longer, Lighter, London, UK, September 2011

Pompeu-Santos, S (2012), *TBM Tunnels of Improved Seismic Behaviour: The TISB Concept.* 15th World Conference on Earthquake Engineering (15WCEE), Lisbon, Portugal, September 2012

Pompeu-Santos, S (2013a), *Inovações em Túneis Rodoviários Realizados com a Técnica TBM (Innovations in Road Tunnels Built with the TBM Technique).* 7th Portuguese Road Congress (7CRP) (in Portuguese), Lisbon, Portugal, April 2013

Pompeu-Santos, S (2013b), *Contributos para a Melhoria do Sistema de Mobilidade e Acessibilidade da Cidade de Lisboa (Contributions for the Improvement of the Mobility and Accessibility of the City of Lisbon).* Lisbon Geographic Society (SGL), Memory n.15 (in Portuguese), September 2013

Pompeu-Santos, S (2014), *New Trends in the Tunnels of the Future.* IABSE Symposium Engineering for Progress Nature and People, Madrid, Spain, September 2014

Pompeu-Santos, S (2016), *Sustainable TBM Tunnels for Tomorrow.* Second International Conference on Concrete Sustainability (ICCS2016), Madrid, Spain, June 2016

Pompeu-Santos, S (2018), *The Long Traffic Tunnels of Tomorrow.* IABSE Symposium Tomorrow's Megastructures, Nantes, France, September 2018

Pompeu-Santos, S (2019a), *Grandes Projetos de Obras Públicas. Desafios Portugal 2030 (Major Public Works Projects. Challenges Portugal 2030).* Scribe (in Portuguese), Lisbon, March 2019

Pompeu-Santos, S (2019b), *The Long Rail and Road Tunnels of Tomorrow.* FIB Symposium Concrete- Innovations in Materials, Design and Structures, Krakow, Poland, May 2019

Pompeu-Santos, S (2019c), *A Travessia do Tejo em Lisboa. Passado, Presente e Futuro (The Tagus Crossing in Lisbon. Past, Present and Future).* Scribe (in Portuguese), Lisbon, September 2019

Pompeu-Santos, S (2020), *Innovative Concepts in TBM Tunnels.* Tunnel Engineering – Selected Topics (Chapter 7), IntechOpen, UK, April 2020

REFER (1999), *A Ponte 25 de Abril. Construção da 2ª Fase- Instalação do Caminho-de-Ferro.* REFER (in Portuguese), 1999

Chapter 9

TBM tunnel solution for a large mixed rail and road crossing based on the TMG and TMF concepts

Abstract

This chapter presents the Fehmarnbelt Fixed Link project, a large mixed railway and road crossing in the Baltic Sea, as well as three alternative approaches for its realization: the Immersed tunnel solution, a conventional TBM tunnel solution and an alternative TBM tunnel solution, developed by the author based on the TMG and TMF concepts.

Detailed descriptions of the main aspects of each of the solutions are presented, which are compared on basis of the most relevant parameters for the project.

The advantages of the alternative TBM tunnel solution based on the TMG and TMF concepts are shown.

9.1 INTRODUCTION

The Fehmarnbelt Fixed Link is a Danish-German project, between Denmark and German, in the Baltic Sea, 18 km long, to provide a direct link by rail and road between the two countries, allowing to connect Scandinavia to Central Europe.

The project is part of the expansion of the trans-European transport network of the European Union (TEN-T), being co-financed by EU funds. It will be owned and financed by Denmark and to be repaid by the users. It is being managed by Femern A/S, a subsidiary of the Danish state-owned Sund & Bælt Holding A/S.

It is a very challenging project, the largest mixed rail and road crossing ever built. The road crossing will also be the second longest road tunnel in the world (after the Laerdal Tunnel in Norway, as mentioned), and the longest highway tunnel.

Studies for the project started in the middle of the 1990's. It has been studied several variants by the Femern A/S consultants, starting with a suspension bridge solution, followed by a cable-stayed bridge solution.

As both bridge solutions received much opposition, mostly from environmental organisations, an immersed tunnel solution was further studied. Later, by 2010–2011, a solution with TBM tunnels was also studied.

However, based on a recommendation by Femern A/S issued on November 2010, on January 2011 the Danish government took a preliminary decision to adopt an immersed tunnel on the link, with an estimated cost of EUR 5500 million.

DOI 10.1201/9781003144861-10

This recommendation was based on a comparison between the cable-stayed bridge and the immersed tunnel. Although the estimated costs of the two solutions were quite similar, the immersed tunnel was chosen because it is considered to have less environmental impact.

As for the TBM tunnel solution, although considered to have less environmental impact than the Immersed tunnel solution, was rejected because the estimated cost was higher.

Meanwhile, in late 2011, Femern A/S published a "Consolidated Technical Report" (Femern 2011), dated December 2011, with descriptions of the variants studied so far. From this report, the author concluded that the TBM tunnel solution studied was not optimized and a better approach in TBM tunnels could be achieved for the project.

Thus, the Author developed an alternative TBM tunnel solution, based on the TMG and TMF concepts (EPO 2011; INPI 2009a, 2009b), which report (Pompeu-Santos 2012a), was sent to the Danish authorities in February 2012, that was shown to be much more cost-effective than the "official" TBM tunnel solution and the Immersed tunnel solution. However, despite the exchange of subsequent messages and explanations (Pompeu-Santos 2012b), the alternative proposal was not accepted.

In the alternative TBM tunnel solution the link is formed of two tubes, one for railway traffic and the other for road traffic, both provided with separated and isolated traffic directions and adopting innovative and very reliable safety concepts. Thus, the alternative TBM tunnel solution keeps the environmental advantages that characterize TBM tunnels, adding a higher level of safety during operation and at a much lower cost than any of the "official" solutions.

In middle 2013, the Immersed tunnel solution was then submitted to public consultation of the EIA (Environmental Impact Assessment) in Denmark. The author participated on the consultation with an updated version of his proposal (Pompeu-Santos 2013d). Despite criticism from several participants (including the author), the project got its approval.

The Consultation report (TM, 2014), however, does not comment on the alternative TBM tunnel solution presented by the author, mentioning only that "a TBM tunnel solution had already been analysed by the Danish authorities and considered less favourable than the Immersed tunnel solution". Now, as will be shown below, the alternative TBM tunnel solution proposed by the author is completely different from that developed by the Femern A/S consultants.

Subsequently, in 2014, The EIA of the project was submitted to public consultation in the state of Schleswig-Holstein, in Germany, where, after a rather tough process, mainly due to opposition from environmental organisations, approval was also granted in February 2019.

The construction works have since been awarded. However, the future of the project in its current form is still uncertain, as some opponents announced that they would appeal the decision to the German courts.

The financial viability of the project is also doubtful, however the payment period has been extended to 36 years. It should also be noted that Scandlines, the operator of the existing ferry line, has announced that it will maintain its service when the link goes in operation.

9.2 THE FEHMARNBELT FIXED LINK

The Fehmarnbelt Fixed Link aims to connect the Lolland Island, in Denmark, to the Fehmarn Island, in Germany, through the Fehmarn Belt. It will constitute the shortest way to connect Scandinavia to the European continent, in continuation of the Oresund Link (figure 9.1) (Femern 2011).

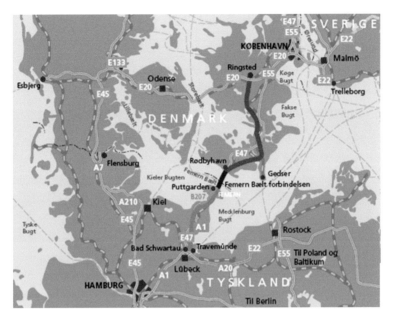

Figure 9.1. Location of the Fehmarnbelt fixed link

The crossing will be for mixed traffic, with two road galleries with two lanes each and two railway galleries, for trains at speeds up to 200 km/h.

The geological profile along the tunnel alignment is shown in figure 9.2 (Femern 2011).

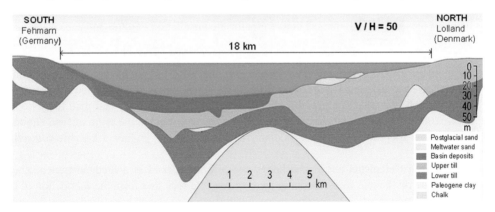

Figure 9.2. Geotechnical conditions of the site

Both sides present smooth slopes near shore areas. Under the seabed the soil comprises an upper Quaternary layer of post and late glacial deposits (clay and silts) followed by a Paleogene layer of highly plastic clay. The German side is characterized by Paleogene clay and some clay-till, the central basin by sand silts and clays, while the Danish side is dominated by thick deposits of clay-till. At its deepest part the water height is about 34m.

9.3   THE IMMERSED TUNNEL SOLUTION

9.3.1   General description

The Immersed tunnel solution is a conventional immersed tunnel, consisting of a single prismatic tube 18 km long, 42.2m wide and 8.9m high (figure 9.3) (Femern 2011).

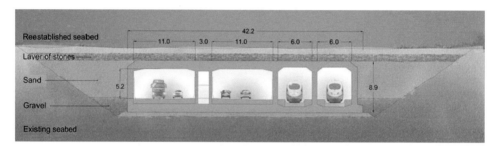

Figure 9.3.   The Immersed tunnel solution. Current cross-section

It is formed by 89 large precast concrete elements, the standard 217m long. At intervals of about 1800 meters, there are special elements that penetrate deeper into the ground, 39m long, 45.0m wide and 13.1m high, creating a lower level to house equipment and to provide connection between the galleries. There also are cut-and-cover sections, about 0.5 km long, at both ends of the tunnel.

The tunnel is provided with four traffic galleries separated by separating walls: two road galleries, 11.00m wide each (each with two lanes) and 5.20m high; and two (ballastless) railway galleries, 6.00 m wide and 6.00m high. It also includes a service gallery, placed between the two road galleries, 2.00m wide, for the installation of pipes and cables and to be used as temporary refugee although not allowing to be used by vehicles (figure 9.3).

The railway galleries are provided with emergency walkways 1.30m wide, while the road galleries are provided with emergency lane and inner edge but not having emergency walk-ways. There will be escape doors every 110m in the separating walls (Andersen 2017; Femern 2011), much less what is allowed by the EU regulations for railway and road tunnels (EU 2004, 2008). The structural elements are designed to withstand in the event of fire. The traffic galleries are provided with longitudinal ventilation through jet fans, which are controlled manually.

The precast elements are placed below the original seabed in a dredged trench, about 100m wide on the top, over a bedding layer of crushed rock that form the foundation of the elements. A combination of locking gravel fill and sand fill is then backfilled along the sides of the elements, while a protection layer of stones, 1.20m thick, is placed across the top of the elements. Part of the dredged material is to be placed on the protective layer in order to re-establish the natural seabed.

It is assumed that the execution of the works presents some risks, since they are developed at the surface of the open sea, in an area of intense ship traffic and using precast elements significantly larger than those used in previous projects.

9.3.2 Environmental impact

As generally recognised, the environmental impact of an immersed tunnel in the site of the project is very significant.

Among others, it is worth noting the large area of natural seabed of the German Natura 2000 site that will be disturbed by the construction works, in a width of over 100 meters along the entire length of the tunnel.

Furthermore, the dredged material not re-used on the back-fillings of the elements would, in principle, have to be decontaminated before being placed in-shore, because it is "superficial" soil, taken from the seabed, in an area of intense ship traffic. It is also likely find remains of explosives and unexploded bombs of the WWII during the excavation of the ground surface of the sea.

It is also to be noted the huge volume of excess dredged material that will have to be placed in reclamation areas: 14.8 million cubic meters (Femern 2011).

Also impressive is its significant "foot print", with the following quantities of the most representative materials used: concrete- 3.0 million cubic meters, rock- 3.1 million cubic meters, and sand- 5.1 million cubic meters (Femern 2011).

9.3.3 Safety in operation

The safety concept of the immersed tunnel solution is based on conventional vehicles (passenger buses and fire trucks) that will access through the road galleries. It is not known whether these vehicles enter the tunnel through one or both the tunnel portals.

In the event of an accident or fire inside one of the traffic galleries, users will leave this gallery through the escape doors on the separating walls, which are spaced 110 meters apart, and arrive to the nearest road gallery, where they will wait for the rescue vehicles (Femern 2011).

However, despite the insistence of Femern A/S that this is a great asset of the solution (Andersen 2017; Femern 2013b), as shown in Chapter 3, this reduced distance between the escape doors, well below the 500m allowed by EU rules (EU 2004, 2008), does not mean a high level of safety for tunnel users.

On the contrary, the safety concept of the tunnel presents several significant shortcomings (Pompeu-Santos 2014a), namely (figure 9.4) (Femern 2013b):

– The escape doors are of the sliding type, which are slow to respond, which can lead to risky situations;
– The escape doors open directly onto the adjacent traffic gallery, so in the event of a fire, there is a risk of smoke entering the refuge gallery when the escape doors are opened;

Figure 9.4.   The Immersed tunnel solution. Rescue system

- The road galleries do not have emergency walkways, so users of those galleries escape directly to the road lanes, thus with the panic installed among users, those behind will push those in front, which can also lead to risky situations;
- The arrival of dozens, or even hundreds of people (in the case of a train) to the nearest road gallery, fleeing from an incident gallery, risks interrupting the flow of traffic in that gallery, impeding the arrival of the emergency vehicles (rescue vehicles and fire trucks);
- The rescue of the passengers of a train with 400 people, for example, will require about a dozen buses, which may take several hours to have them at the site of the incident, unless the buses with their drivers are permanently parked at the tunnel portals;
- The traffic flow in the non-incident galleries will be significantly disturbed by the occurrence of any safety problem in one of the traffic galleries of the tunnel.

Finally, a very serious question: the escape of users from the outer railway gallery. Users have to cross the inner railway gallery to reach the nearest road gallery (figure 9.4), which will be very dangerous and therefore should not be acceptable.

In fact, this tunnel concept is an extrapolation of the basic concept of immersed tunnels, which are made up two traffic galleries of the same type, with a service gallery in the middle. Joining road and rail traffic in the same immersed tunnel in this way does not work; the two safety systems (safety in operation) must be separated.

Thus, there should be no escape doors in the separating wall between the inner road and the rail galleries, and the access of emergency services and rescue of users in the railway galleries should be done through emergency trains, as in the conventional TBM dual-tube and mono-tube railway tunnels, as in the examples shown in Chapter 4.

9.3.4   Cost

The cost of the tunnel was estimated at EUR 5500 million (Femern 2011), which is similar to the values of the proposals received in the meantime for the construction, which means that it is a realistic estimate.

Given that the project was granted with EUR 600 million in EU funds, the financial effort of the promoter will therefore be around EUR 4900 million, to be repaid over a period of 36 years, as mentioned above.

9.4 THE BASIC TBM TUNNEL SOLUTION

9.4.1 General description

The basic ("official") TBM tunnel solution consists of three parallel tubes; one tube for the railway traffic, and two tubes for the roadway traffic, approximately 20 km long each (figure 9.5) (Femern 2011).

Figure 9.5. The Basic TBM tunnel solution. Current cross section

The railway tube has an inner diameter of 15.20m, and is provided with several internal partition elements, to allocate two parallel and independent ballast-less railway galleries, each for a direction of traffic, and several galleries, for cables and pipes and for access of emergency vehicles for service and rescue of users, located at the middle and bottom of the tunnel. The railway galleries are provided with emergency walkways on the inner side.

The outer wall of the tube will thus be about 0.60m thick, therefore, considering 0.15 m clear space to be injected, the excavated diameter will be 16.70m, a very large diameter for TBM tunnels.

The road tubes have an inner diameter of 14.20m, being also provided with partition elements, each allocating a road gallery on the upper floor, with two traffic lanes, emergency lane and inner edge (but without emergency walkways), and a service gallery on the lower floor, for access of emergency services and rescue of users, with access to vehicles of normal height, as well as several technical galleries for cables and pipes.

The outer wall of the two tubes will thus be about 0.55m thick, therefore, considering also 0.15m of clear space to be injected, the excavated diameter of the tunnel will be 15.60m, also a large diameter.

9.4.2 Environmental impact

Being made up of TBM tunnels, the environmental impact of the solution will not be significant, namely there will be no disturbance of the natural seabed.

Disturbance of natural habitats may occur in-shore only, when placing the excavated material, the effect of which will depend on the boring technology used.

If EPB type machines be used it will not be necessary any treatment, because it is "virgin" material, taken from the interior of the ground. If Slurry or Mixshield type machines be used, it will be necessary to make the separation of the bentonite from the spoil (after its pumping from the head of the excavation), being, however, a common process, nowadays.

9.4.3   Safety in operation

The access of emergency services and the rescue of users in both the railway tunnel and the road tunnels adopt conventional yet complex concepts.

In the event of an accident or fire, users will be rescued using conventional vehicles that use the service gallery below, which they will access to through escape doors and stairs placed inside vertical access galleries, regularly arranged along the tunnels.

9.4.4   Cost

The cost of the basic ("official") TBM tunnel solution was estimated at EUR 6800 million (2008 prices) (Femern, 2011), but it seems to be overestimated.

In fact, as mentioned above, the basic TBM tunnel solution consists of three tubes: one railway tube with an excavated diameter of about 16.70m and two road tunnels with excavated diameters of about 15.60m, which leads to excavated volumes of about 4.38 million cubic meters for the railway tunnel and about 3.82 million cubic meters for each road tunnel. Thus, admitting that the unit costs of railway tunnels are about 70% of the unit costs of road tunnels (Pompeu-Santos 2015), unit costs of EUR 445 per cubic meter for the railway tunnel and EUR 635 per cubic meter for the road tunnel are obtained.

However, compared to the values obtained for similar TBM tunnels (see, Chapter 2), these values are very high, so the estimated cost of the solution should be overestimated by about 10–15%, therefore, it should be in the order of EUR 5900 million.

9.5   COMPARISON BETWEEN THE IMMERSED TUNNEL SOLUTION AND THE BASIC TBM TUNNEL SOLUTION

As mentioned above, on November 2010, Femern A/S recommended to the Danish government the adoption of an Immersed tunnel on the link.

The justification for this recommendation was that, "although the TBM tunnel solution has less environmental impact, the Immersed tunnel solution was preferable because of its lower cost (about 25% less) and shorter time of construction (6.5 versus 8 years)". The Consolidated Report issued later by Femern A/S in December 2011 (Femern 2011) provides more detailed information on this matter.

These statements need to be commented on.

Regarding costs, it should be noted that, as it was shown above, the estimated cost of the basic ("official") TBM tunnel solution is over-estimated in at least 10–15%, so the difference of costs will not be very significant. However, as also mentioned, the basic TBM tunnel solution is not optimized and a better TBM tunnel solution can be achieved.

Regarding the construction time of a TBM tunnel solution, estimated by Femern A/S at eight years, it is also not correct.

In fact, even if only one TBM is used to bore each tunnel, with the excavation rates offered by modern TBMs, of at least 100 meters of tunnel per week on average, the excavation of the tunnels takes less than 4 years. Therefore, considering one year for the delivering of the machines and 1.5 years for the additional complementary works, the construction time will be about 6.5 years, ie the same that of the immersed tunnel solution. If instead, there are TBMs starting at both tunnel portals, the excavation time will be halved, so the construction time of a TBM tunnel solution will be even significantly shorter than that of an immersed tunnel solution.

However, from the beginning of 2013, Femern A/S changed its justifications, having issued several documents (Femern 2013a, 2013b and 2013c), seeking to devaluate any alternatives in TBM tunnels in relation to the immersed tunnel solution.

One of the arguments was to devaluate the importance of the gap in environmental impacts, stating that a bored tunnel will only have a "slight" advantage compared to an immersed tunnel. However, it is well known that the environmental impact of an immersed tunnel is much higher than that of a TBM tunnel. This is indisputable, as it is intrinsic to each of the techniques.

It also starts to emphasize risks associated with the construction of TBM tunnels in the Fehmarnbelt, stating that "the variable conditions of the soil and the likely presence of large boulders could lead to an increase in cost and construction time". It is to be said that, as it will be shown below, large TBM tunnels were built in last two decades in similar conditions or even worst, without significant problems.

Another argument that came to be presented was that the Immersed tunnel was very safe, the "safest in the world" so far (Femern 2013b). However, as it was shown above, its safety concept presents significant risks.

9.6   ALTERNATIVE TBM TUNNEL SOLUTION BASED ON THE TMG AND TMF CONCEPTS

9.6.1   General description

As mentioned above, the TBM tunnel solution developed by the Femern A/S consultants is not optimized and a more advantageous TBM tunnel solution can be achieved.

In the following, an alternative TBM tunnel solution developed by the author, based on the TMG and TMF concepts (EPO 2011; INPI 2009a, 2009b), for railway and road tunnels, respectively, is presented.

The alternative TBM tunnel solution consists of two separate and parallel tunnels, one for railway traffic and the other for road traffic, placed next to each other at a distance of about 15 to 20 meters, going deep into the ground to elevation −63 (water pressure of 6.3 bar), complemented with access ramps at the ends (Buxton 2014; Pompeu-Santos 2012a, 2013a, 2013b, 2013c, 2013d, 2014, 2015, 2016a, 2016b, 2016c, 2017, 2020) (figure 9.6).

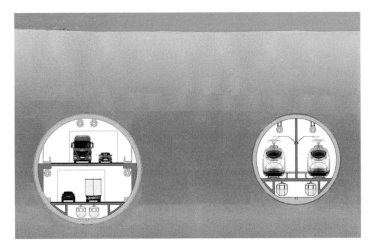

Figure 9.6.   The alternative TBM tunnel solution. Current cross-section

The railway tunnel is about 20 km long and has an inner diameter of 11.50m, using precast segments 0.45m thick (Di/25) and 0.15m clearance to be injected, thus the excavated diameter of the tunnel is 12.70m, a common size for TBM tunnels (figure 9.7).

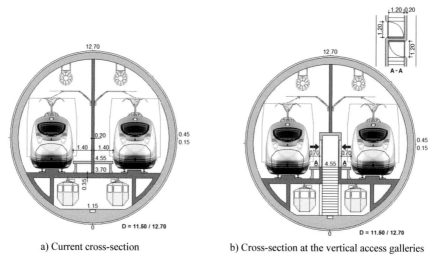

a) Current cross-section   b) Cross-section at the vertical access galleries

Figure 9.7.   The alternative TBM tunnel solution. Cross-section of the railway tunnel

A slab (supported on two intermediate small walls) and a central wall are built, creating two parallel and isolated rail galleries (ballast-less), each one for a direction of traffic, with 38 m^2 cross-sectional area each, and a service (emergency) gallery below, 2.10m high.

The slab is strengthened in the transverse direction through longitudinal ribs placed over the intermediate supports, which also function as anti-derailment devices.

Variant B of the TMG concept (see, Chapter 5) is used; therefore, the emergency walkways, 1.40m wide, are placed inside, along both sides of the separating central wall.

The galleries under the emergency walkways are used as ducts for pipes and cables.

On both emergency walkways there are 49 escape doors of the rotating type, spaced 400m apart (less than the 500m allowed by EU rules), which give access to the service gallery through vertical access galleries (provided with stairs), placed in the middle of the tunnel, along the separating wall. Inclined platform lifts, arranged along the stairs, provide access for people with limited mobility.

Although the placement of the vertical access galleries in the middle of the tunnel causes a slight localised reduction (less than 2 m^2) of the cross-sectional area of the railway galleries in lengths of about 10 meters, the cross-sectional area of these galleries is identical to that of the railway galleries of the immersed tunnel solution, 36 m^2. In addition, as these narrower areas are located outside the portal areas of the tunnel, the comfort conditions inside the trains will not be affected.

The service gallery will be equipped with four EMEVs ("Emergency Monorail Electric Vehicles") "trains", that are parked two at each tunnel portal and circulate suspended from the slab in two parallel lines, to provide local access to emergency personnel and the evacuation of users to out of the tunnel, in the event of an accident or fire inside the tunnel (figure 9.7). Lay-out details of the EMEVs are presented in Chapter 5.

The road tunnel is about 19 km long and has an inner diameter of 14.20m. Using precast segments 0.55m thick (Di/25) and 0.15m clearance to be injected, the excavated diameter of the tunnel is 15.60m (figure 9.8).

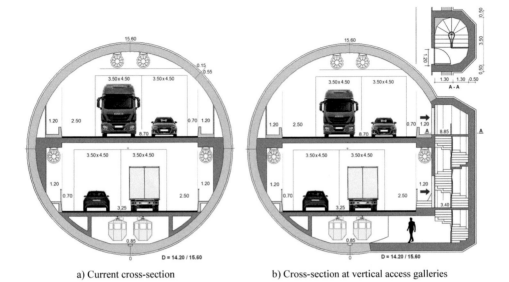

a) Current cross-section   b) Cross-section at vertical access galleries

Figure 9.8.   The alternative TBM tunnel solution. Cross-section of the road tunnel

Although it is a large diameter, as mentioned above (see, Chapter 4), TBM tunnels with larger diameters have already been built.

Inside the TBM tunnel, two intermediate slabs are built, creating three superimposed galleries: two road galleries, independent and isolated, each one for a traffic direction, 5.00m high and 4.50m clear height (allowing 0.50m for the signaling installation), and a service (emergency) gallery, below, 2.00m high.

Both road galleries have two 3.50m wide lanes, 2.50m wide emergency lane and 0.70m wide inner edge, making a total road width of 10.20m. Thus, the tunnel presents an efficiency index with an impressive value of 0.57, which is 1.35 times higher than the average efficiency indexes of noteworthy examples of dual-tube and mono-tube dual-deck road tunnels, presented in Chapter 4, which is 0.42.

There are also emergency walkways 1.20m wide on both sides of the road galleries, making a total width of 12.60m, wider than the roadway galleries of the Immersed tunnel solution, which is 11.00 m.

Laterally to the emergency walkways, on one side of the tunnel, there are 47 escape doors of the rotating type, spaced 395m apart (less than that the 500m allowed by EU rules), which give access to the service gallery bellow through vertical access galleries, disposed parallel to the tunnel axis and provided with stairs (figure 9.8). Inclined platform lifts disposed along the stairs also provide access for people with limited mobility.

The galleries along the emergency walkways of the upper floor are used as ducts to accommodate pipes and cables.

Inside the service gallery, also four EMEV "trains" will provide access to the emergency personnel and the evacuation of users to outside of the tunnel, in the event of an accident or fire inside the tunnel.

Vertical access galleries are built by locally disassembling precast segments of the outer wall (in lengths of 4.50m, i.e. two rings) and casting new reinforced concrete walls in-situ (see, Chapter 5).

The access ramps at the ends of the tunnel have a special lay-out to make the transition between the superimposed road galleries and the lanes placed side by side of the highway (see, Chapter 5).

On the Lolland side, embankments carried out with soil taken from the excavations, will be placed along the alignments of both tunnels, which allows to reduce their length (figure 9.9).

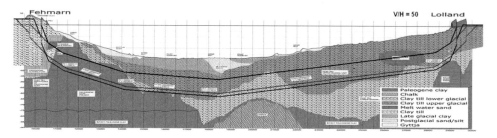

Figure 9.9.   The alternative TBM tunnel solution. Longitudinal sections of the tunnels

9.6.2   Environmental impact

Being constituted of bored tunnels, the alternative TBM tunnel solution will not provoke any disturbance of the natural seabed along the tunnel alignment.

As mentioned above, the disturbance of natural habitats can only occur in-shore, when placing the bored material. However, if EPB type machines are used, no treatment is necessary, as it is "virgin" material, taken from the interior of the ground. If Slurry or Mixshield type machines are used, it will be necessary to separate the bentonite from the soil, which is a common process today.

However, the volume of excavated material that will be placed in the reclamation areas is relatively low, only about 6.2 million cubic meters.

The spending of natural resources on the main building materials ("footprint") is as follows: concrete- 1.9 million cubic meters; and rock and sand, non-significant (Pompeu-Santos 2013d).

9.6.3   Technical systems

Drainage

Both tunnels are equipped with drainage systems, to drain any water entering them, through the joints or from the fire fighting system.

Thus, both traffic galleries of each tunnel are provided with a line of sink drains, regularly distributed along its length, with the effluent being discharged into pipes placed along the underside of the slabs, from which it is conducted for wells created along the bottom of the tunnels (see, Chapter 5, figures 5.20 and 5.21).

From the wells, the effluent is then pumped out of the tunnels through pipes placed along one of the the sides of the service gallery.

If there are discharges of fuel or other contaminating liquids in the pavement of the traffic galleries, it will be possible to remove them from the wells.

Ventilation

In both tunnels the traffic galleries are equipped with longitudinal ventilation through jet fans, which push air and smoke in the direction of the traffic, prepared to act in two scenarios: normal ventilation and fire ventilation (see, figures 9.7 and 9.8).

The service gallery in both tunnels is permanently pressurized through ventilators located at the ends of the tunnels.

In the road tunnel, at the exits of the portals, the polluted air will be taken to treatment plants, placed next to them (see, figures 8.15 and 9.11).

Fire fighting

For fire fighting, the traffic galleries of both tunnels are equipped with active systems of detection and attack, acting jointly.

Heat sensors and smoke detection systems automatically activate WFSs ("Water Based Fire Fighting System") formed of high pressure water mist nozzles, regularly distributed along the tunnel, which are grouped into sections about 30 meters long (see, figures 5.22b and 5.23).

In the railway tunnel, the nozzles are distributed along three "emergency stations" about 750m long each and spaced about 5 km, while in the road tunnel, nozzles are distributed over the entire length of the he tunnel.

After the WFS action, fire brigades that access the site through the service gallery, will then easily extinguish the fire.

Hydrants placed at regular intervals along the emergency walkways of the traffic galleries allow fire fighters to access the pressurized water network.

Fire lines disposed along the traffic galleries (under the emergency walkways of the railway tunnel and along the gallery arranged next to one of the emergency walkways of the upper gallery on the road tunnel) supply pressurized water to the WFSs, as well to the hydrants.

9.6.4 Safety in operation

Being based on TMG and TMF concepts, both tunnels are provided with innovative concepts for safety in operation, which represent a step forward in tunnel safety.

In the event of an accident or fire inside one of the traffic galleries of the tunnels, users will leave that gallery walking along the respective emergency walkway to the nearest escape door, from which they access the service (emergency) gallery below, downing the stairs inside the vertical access galleries.

Inside the service gallery of each tunnel, EMEV ("Emergency Monorail Electric Vehicles") "trains" will provide access for the emergency services (rescue personnel and fire fighters) to the site of the incident and the rescue of users to out of the tunnel (see, figure 5.26).

In the portal areas, access buildings equipped with elevators and stairs provide access of users to the surface (figures 9.10 and 9.11).

Front View   **Longitudinal Section**

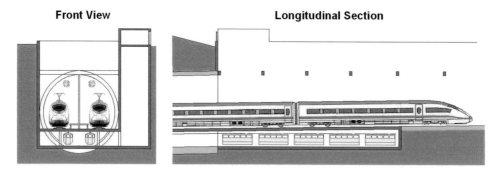

Figure 9.10.   The alternative TBM tunnel solution. Portals of the railway tunnel

EMEV "trains" consist, in principle, of five units in the case of the railway tunnel (total capacity of 100 people each) and two units in the case of the road tunnel (total capacity of 40 people each). In both tunnels, the four EMEV "trains" are parked two on each portal.

Front view   **Longitudinal section**

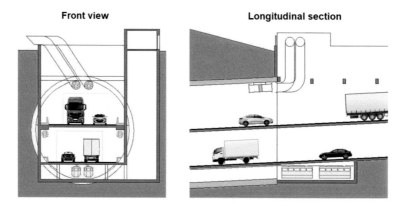

Figure 9.11.   The alternative TBM tunnel solution. Portals of the road tunnel

In the event of an accident or fire inside one of the traffic galleries in the railway tunnel (which is the most demanding), it will take less than 30 minutes to rescue all the occupants of a train with a capacity of 400 passengers if it stops inside the tunnel.

In both tunnels in the event of an incident situation in one of the traffic galleries, there will be no disruption to traffic flow in the non-incident gallery.

Inside the railway tunnel, at about half of its length, a cross-over will be installed, thus, a lengthy opening will be created in the separating wall, protected with a fireproof sliding door, to allow trains to switch tracks, in case of need (maintenance works, broken train, etc.).

9.6.5 Cost

The cost of the alternative TBM tunnel solution was estimated based on the unit costs obtained for the basic TBM tunnel solution, which are EUR 445 and EUR 635 per cubic meter of excavation for the railway and the road tunnels, respectively (see, Section 9.4.4).

Thus, for the alternative TBM tunnel unit costs of EUR 445 per cubic meter of excavation for the railway tunnel (the same as that obtained for the basic TBM tunnel solution) and EUR 650 per cubic meter of excavation for the road tunnel (slightly higher than that obtained for the basic TBM tunnel solution, as there are vertical access galleries to build) were adopted.

Considering that, in this case, excavated volumes are about 2.54 million cubic meters for the railway tunnel and 3.63 million cubic meters for the road tunnel, the estimated costs will be EUR 1150 million and EUR 2400 million for the railway and the road tunnels, respectively, therefore, the estimated cost of the alternative TBM tunnel solution will be EUR 3550 million (Pompeu-Santos 2013d),

However, as mentioned above, it seems that the estimated cost of the "official" TBM tunnel solution is overestimated by about 10–15% (Pompeu-Santos 2017), so the estimated cost of the alternative TBM tunnel solution should be in the order of EUR 3100 million, which is less than two thirds the estimated cost of the Immersed tunnel solution.

9.7   COMPARISON BETWEEN THE IMMERSED TUNNEL SOLUTION AND THE ALTERNATIVE TBM TUNNEL SOLUTION

9.7.1   Introduction

A comprehensive comparison between the alternative TBM tunnel solution and the Immersed tunnel solution is carried out on basis of the most significant parameters for this project: environmental impact, technical risks during construction, proven technology, safety and emergency response during operation, and schedule and finances.

9.7.2   Environmental impact

With regard the Immersed tunnel solution, in addition to extensive disturbances in the areas of the portals, there will be disturbance of the natural seabed along the entire length of the tunnel, in a width of at least 100m, namely on the German Natura 2000 site, located in the middle of the Fehmarnbelt. This is a fact that cannot be disguised, because it is intrinsic to the construction of immersed tunnels.

There will also be disturbances in natural habitats in the reclamation areas, for the placement of the great quantity of dredged material not re-used in the back-fillings. In addition, this material should, in principle, be decontaminated before placing, as it is "superficial" soil, taken from the seabed, in an area of intense ship traffic.

As regard the alternative TBM tunnel solution there will be no disturbance of the natural seabed along the tunnel. Disturbances of natural habitats may only occur in-shore, when placing the bored material in the reclamation areas, which will depend on the boring technology used.

If EPB type machines are used, no treatment of the bored material is necessary, because it is "virgin" material, taken from the interior of the ground. If Slurry or Mixshield type machines are used, it will be necessary to separate the bentonite from the excavated soil, but that is a common and very reliable process today.

It should also be noted that in the alternative TBM tunnel solution the volume of material to be placed in the reclamation areas is less than half that of the Immersed tunnel solution: 6.2 million m^3 versus 14.8 million m^3 (Femern 2011).

Regarding the natural resources used ("footprint), on Table 9.1 the quantities of the most representative building materials required in the alternative TBM tunnel solution and the Immersed tunnel solution (Femern 2011) are shown.

Table 9.1. Resources required

	Alternative TBM tunnel solution	Immersed tunnel solution
Concrete (m^3)	1.9 million	3.0 million
Rock (m^3)	Non-significant	3.1 million
Sand (m^3)	Non-significant	5.1 million

It can be seen the huge consumption of primary building materials in the Immersed tunnel solution compared to the alternative TBM tunnel solution.

Another advantage of the alternative TBM tunnel solution is the less severe impact on the area, due to the smaller site for construction activities and production facilities.

In summary, it is concluded that the environmental impact of the alternative TBM tunnel solution is much less than that of the Immersed tunnel solution.

9.7.3 Technical risks during construction

Regarding risks during construction, the two solutions present quite different situations.

In which concerns the alternative TBM tunnel solution, it should be emphasized that to build bored tunnels in the Fehmarbelt ground does not pose a particular high risk. The main risks during construction are those associated with the execution of the vertical access galleries in the road tunnel. However, these risks should not be over-estimated.

In fact, as mentioned in Chapter 5, in situations where there is water pressure around the tunnel, injections of jet-grouting will create a kind of "concrete cake", within which the works will be carried out safely, without leading to a significant increase in cost.

As also mentioned, in the railway tunnel, the vertical access galleries are placed at the middle of the tunnel, on the separating wall, so their construction do not represent any particular risk.

Regarding the navigation safety, the construction of the alternative TBM tunnel solution, will not cause any affection on the ship traffic in the Fehmarnbelt.

In which concerns the Immersed tunnel solution significant risks during construction and for the navigation safety should be considered, since most works will be carried out at the surface of the open sea, in an area of intense ship traffic (more than 50 thousand vessels a year). These risks will still be exacerbated by the use of precast elements that are significantly larger than those used in previous projects.

In summary, despite the difficulty in comparing such different situations, it can be concluded that, regarding the risks associated with construction, neither alternative has a significant advantage over the other.

9.7.4   Proven technology

In the alternative TBM tunnel solution, both tunnels use proven technology, since identical tunnels, under similar (or even worse) conditions were already built.

Regarding the railway tunnel, TBM tunnels enclosing two separate traffic galleries (mono-tube dual-galleries concept), there is the very successful example of the Groenehart tunnel, in The Netherlands, in the high-speed line HSL-Zuid, 7 km long, with an excavated diameter of 14.87m (see, Chapter 4). The railway tunnel of the alternative TBM tunnel solution has a smaller diameter, only 12.70m. Based on the dual-tube tunnel concept many railway tunnels exist, however with smaller diameters.

Regarding the road tunnel, as mentioned, TBM tunnels with even larger diameters than the 15.60m of the proposed tunnel have already been built, such as the Santa Lucia Tunnel, near Firenze (Italy), a single tube 15.87m diameter with 3 lanes, the SR99 Tunnel, in Seattle (USA), a mono-tube dual-gallery 17.50m diameter with two lanes in each direction, and the Northern part of the Tuen Mun-Chek Lap Tunnel in Hong Kong (China), a mono-tube 17.60m in diameter with three lanes. Still more impressive, the Orlosvski Tunnel, a mono-tube dual gallery 19.25m diameter with three lanes in each direction, in Saint Petersburg (Russia), however it was not built.

As for soil conditions, as mentioned in Chapter 4, examples of TBM tunnels built in similar situations, also exist, such as the Westerschelde tunnel (in Netherlands, not far from the Fehamernbelt), 7 km long, built under water pressures up to 6 bar, or the Shanghai Changjiang tunnel (in China), 7.5 km long, in heterogeneous soil and water pressure up to 6.5 bar, both formed by twin-tubes connected by cross-passages.

With regard to the Immersed tunnel solution, this type of tunnels have been used in very few situations. One can mention the Drodgen (road and rail) Tunnel, on the Oresund Link, between Denmark and Sweden, 3.5 km long, the road Busan-Geoje Tunnel, in South Korea, 3.2 km long, the Marmaray metro tunnel, on the Bosphorus, in Istambul, Turkey, 1.4 km long, and in the 6 km long undersea section (navigation channel) of the Hong Kong–Zhuhai–Macau Bridge, in China. However, the experience with the immersed tunnel in Istambul was not very positive, thus, for the construction of the Eurasia Tunnel, close by, as mentioned in Chapter 4, a TBM tunnel was chosen. The same happened in Hong Kong, where, for the construction of the undersea Tuen Mun-Chek Lap Kok Link, a TBM tunnel was eventually adopted.

Anyway, it can also be considered that the immersed tunnel is proven technology for this project, although it must be taken into account that the construction site is an area of intense ship traffic and the precast elements are significantly larger than those used in the aforementioned tunnels.

In summary, it may be concluded that, in this respect, neither alternative has a significant advantage over the other.

9.7.5   Safety in operation

As mentioned in Chapter 3, EU rules (EU 2004, 2008) require that traffic galleries in railway and road tunnels must be provided with escape doors spaced 500m maximum. Therefore, it is assumed that this distance between escape doors represent an acceptable risk for users and can be used safely.

Thus, it has to be played down the importance of the difference in distance between escape doors, which is 110m in the Immersed tunnel solution and 400m in the tunnels of the alternative TBM tunnel solution. In fact, as shown in Chapter 3, for these distance levels, the difference between the distance values is hardly relevant.

Safety in operation of the Immersed tunnel solution
In the Immersed tunnel solution the rescue of users relies on conventional vehicles that will use the road galleries of the tunnel, to which users of the road and rail galleries will have access to through the escape doors arranged on the separating walls of the tunnel.

As mentioned in Section 9.3.3, this rescue concept presents significant dangers, with unacceptable risks for the safety of users (see, figure 9.4).

In fact, this type of rescue may be suitable for users of the road galleries, who leave the incident gallery and move on to the other road gallery (crossing the service gallery). However, the road galleries do not have emergency walkways, so it will be dangerous for users to wait for the arrival of the rescue vehicles, inside these galleries.

It may also appear suitable for users to escape from the inner railway gallery to the inner road gallery, despite the difficulty of accommodating several hundred passengers of an incident train at the emergency lane of the inner road gallery, awaiting the rescue vehicles.

A really serious problem is how to escape from the outer railway gallery. In fact, users have cross the inner railway gallery to reach the inner road gallery, which will be very dangerous, thus the tunnel should not be considered safe.

In addition, the traffic flow in the non-incident galleries of the tunnel will be significantly disrupted by the occurrence of any safety problem within one of the traffic galleries of the tunnel.

Safety in operation of the alternative TBM tunnel solution
In the alternative TBM tunnel solution, the traffic galleries of both tunnels are equipped with emergency walkways along their entire length, through which users can escape to the service gallery below (via the vertical access galleries), walking to the nearest escape door, which are arranged every 400 m, less than the maximum allowed by EU rules (EU 2004, 2008). In addition, in the railway tunnel, escape doors (and vertical access galleries) with less spacing can, however, be easily adopted, without significant increase in costs.

Within the service gallery of each tunnel, dedicated EMEVs ("Emergency Monorail Electric Vehicles"), which are parked at both tunnel portals and circulate suspended from the ceiling slab, on two parallel lines, provide local access to emergency personnel (rescue personnel and fire fighters) and the evacuation of users to out of the tunnel in the event of an accident or fire inside the tunnel (see, Chapter 5, figure 5.26).

In the railway tunnel (which is the most demanding) four EMEV "trains" of, in principle, five units each, therefore with capacity for at least 100 people each, which are parked two in each portal (see, figure 9.10), will take less than 30 minutes to rescue all the occupants on passenger trains if they stop inside the tunnel.

As in each tunnel the rescue system does not make use of the next traffic gallery, the traffic flow in the non-incident gallery of the tunnel will not be affected by these events.

In summary, it is concluded that the Immersed tunnel solution presents significant dangers in its safety concept and should not be considered safe. In contrast, the alternative TBM tunnel solution presents an innovative and very reliable safety concept, which represents a step forward for tunnel safety.

9.7.6   Schedule and finances

Regarding the time schedule, as mentioned in Section 9.5, it can be considered that the construction times of the alternative TBM tunnel solution and the Immersed tunnel solution will be identical, 6.5 years each.

As regard the construction cost, that of the Immersed tunnel solution was "officially" estimated at EUR 5500 million (2008 prices).

In the case of the alternative TBM tunnel solution, the cost estimated on basis of the unit costs of the tunnels of the "official" TBM tunnel solution (see Section 9.6.7), lead to an estimated cost of EUR 3550 million.

However, as mentioned above, the estimated cost of the "official" TBM tunnel solution seems to be overestimated by about 10–15%, so the estimated cost of alternative TBM tunnel solution will be of the order of EUR 3100 million, which is less than two thirds the cost of the Immersed tunnel solution.

Since the project was granted EUR 600 million in EU funds, this means that with the alternative TBM tunnel solution the financial effort of the promoter will be just over half of that of the Immersed tunnel solution.

9.8   CONCLUSIONS

The Fehmarbelt Fixed Link project, a 20 km long mix rail and road crossing in Baltic Sea, between Lolland Island (in Denmark) and Fehmarn Island (in Germany), is in preparation, under the responsibility of Femern A/S, a state-owned Danish company.

Based on a recommendation from Femern A/S, the Danish government decided to adopt an Immersed tunnel on the link. However, an alternative TBM tunnel solution, formed of two tunnels, one for rail traffic and the other for road traffic, based on TMG and TMF concepts, shows to be more advantageous.

In fact, comparing the alternative TBM tunnel solution with the Immersed tunnel solution, although in some parameters there is no a clear advantage of one solution in relation over the other, in significant parameters, such as environmental impact, safety in operation and cost, the alternative TBM tunnel solution shows to be much more favourable.

It is also worth mentioning the relatively small diameter of the railway tunnel and the impressive efficiency index of the road tunnel, 0.57, which is 1.35 times higher than the average efficiency indexes of impressive dual-tube and the mono-tube dual-deck road tunnels presented in Chapter 4.

Regarding the environmental impact, in the alternative TBM tunnel solution it is much smaller than in the Immersed tunnel solution. This is unquestionable, because it is intrinsic to each of the tunnelling techniques.

In particular, while in the alternative TBM tunnel solution the seabed is not be touched, in the Immersed tunnel solution there will be disturbance of the natural seabed along the entire length of the tunnel, in a width of at least 100m, in the German Natura 2000 site.

With regard to safety during operation, the Immersed tunnel solution has significant weaknesses in its safety concept, and should not be considered safe. Among other dangers, the rescue of passengers from trains circulating in the outer railway gallery obliges users to cross the inner railway gallery, which should not be acceptable.

In contrast, the alternative TBM tunnel solution presents innovative and very reliable safety concepts, which represent a step forward for the tunnel safety. The rescue of users is carried out using dedicated EMEVs, remote controlled electric vehicles that circulate inside the service (emergency) gallery, therefore, completely independent of traffic conditions inside the traffic galleries.

As for costs, the alternative TBM tunnel solution has a much lower cost than the immersed tunnel solution, EUR 5500 million, when the estimated cost of the alternative TBM tunnel solution is in the order of EUR 3100 million, which is less than two thirds of the cost of the Immersed tunnel solution.

In conclusion, compared to the Immersed tunnel solution, the alternative TBM tunnel solution based on TMG and TMF concepts is much more cost-effective, having a much lower environmental impact, providing much higher safety during operation and having much lower cost.

REFERENCES

Andersen K S et al. (2017), *Fehmarnbelt Fixed Link: Safety Concepts and Strategies for a Combined Road and Railway Tunnel – Legal Aspects, Ventilation Concept, User Safety, Accessibility for Emergency Services, Evacuation Scenarios*. STUVA Conference 2017, Stuttgart, Germany, December 2017

Buxton L (2014), *Flexible Solutions for Fehmarnbelt?* World Tunnelling Magazine, July– August 2014

EPO (2011), *Tunnel Multi-Storey*. EPO, European Patent EP 2317074A1, November 2011

EU (2004), *EU Directive 2004/54CE: Safety in Roadway Tunnels*. EU, April 2004

EU (2008), *Safety in Railway Tunnels*, EU Decision 2008/163CE, January 2008

Femern (2011), *Fehmarnbelt Fixed Link- Consolidated Report*, Femern A/S, Copenhagen, Denmark, December 2011

Femern (2013a), *Fehmarnbelt Fixed Link. Transboundary Environmental Impact Assessment*. Femern A/S, Copenhagen, Denmark, June 2013

Femern (2013b), *A Safe Fehmarnbelt Tunnel*. Project Information, Femern A/S, Copenhagen, Denmark, June 2013

Femern (2013c), *The Selection of the Immersed Tunnel as the Preferred Solution*. Memo Femern A/S, Copenhagen, Denmark, September 2013

Fixed Link, Ministry of Public Works, Copenhagen, Denmark, September 2013

INPI (2009a), *Túnel de Galeria Múltipla (Tunnel Multi Gallery)*. INPI, Portuguese Patent n. 103769 (in Portuguese), January 2009

INPI (2009b), *Túnel Multi Piso (Tunnel Multi Floor)*. INPI, Portuguese Patent n. 103748 (in Portuguese), August 2009

Pompeu-Santos S (2012a), *Alternative Tunnel Solution for the Fehmarnbelt Fixed Link Based on the TMG and TMF Concepts*. Author's edition, Lisbon, Portugal, February 2012

Pompeu-Santos S (2012b) *Fehmarnbelt Fixed Link. Response to the Technical Note from RAT (Ramboll/ARUP/TEC) dated 2012/08/30 on an Alternative Bored Tunnel (TBM) Solution.* Author's edition, Lisbon, Portugal, December 2012

Pompeu-Santos S (2013a), *Alternative TBM Tunnel Solution for the Fehmarnbelt Fixed Link.* Author's edition, Lisbon, Portugal, March 2013

Pompeu-Santos S (2013b), *Tunnel Solution for the Fehmarnbelt Fixed Link.* Author's edition, Lisbon, Portugal, May 2013

Pompeu-Santos S (2013c), *The TMG and TMF Concepts: The Right Approach for Large Tunnel Crossings.* Word Tunnel Congress (WTC2013) Underground, the Way to the Future, Geneve, Switzerland, June 2013

Pompeu-Santos, S (2013d), *Fehmarnbelt Fixed Link- Alternative TBM Tunnel Solution Based on TMG and TMF Concepts versus the Immersed Tunnel Solution.* Public Consultation in Denmark of the EIA of the Fehmarnbelt

Pompeu-Santos (2014), *Safety During Operation of the Fehmarnbelt Fixed Link. The Immersed Tunnel Solution versus an Alternative TBM Tunnel Solution Based on TMG and TMF Concepts.* Author's edition, Lisbon, Portugal, May 2014

Pompeu-Santos S (2015), *Optimized TBM Tunnel Solution for the Fehmarnbelt Fixed Link.* FIB Symposium Concrete- Innovation and Design, Copenhagen, Denmark, May 2015

Pompeu-Santos S (2016a), *Optimized TBM Tunnel Solution for the Fehmarnbelt Fixed Link Based on TMG and TMF Concepts.* SPS Consulting, Lisbon, Portugal, February 2016

Pompeu-Santos S (2016b), *Sustainable TBM Tunnels for Tomorrow.* Second International Conference on Concrete Sustainability (ICCS2016), Madrid, Spain, June 2016

Pompeu-Santos S (2016c), *Meet the Challenges of the Fehmarnbelt Fixed Link.* IABSE Congress Challenges in Design and Construction of an Innovative and Sustainable Built Environment, Stockholm, Sweden, September 2016

Pompeu-Santos S (2017), *Make Feasible the Fehmarnbelt Fixed Link.* FIB Symposium High Tech Concrete- Where Technology and Engineering Meet, Maastricht, Netherlands, May 2017

Pompeu-Santos S (2020), *Innovative Concepts in TBM Tunnels.* Tunnel Engineering – Selected Topics (Chapter 7), IntechOpen, London, UK, April 2020

TM (2014), *Høringsnotat. VVM- Redegørelse for en Fast Forbindelse over Femern Bælt (Consultation Report. EIA Report for a Fixed Link across the Fehmarnbelt).* Transport Ministeriet (in Danish), Copenhagen, Denmark, October 2014

Chapter 10

Conclusions and future perspectives

Abstract

In this chapter the main conclusions drawn above are presented, as well as the future perspectives for TBM tunnels with the application of the TISB, TMG and TMF concepts to
challenging projects, such as the mono-tube TBM road tunnels with three lanes in each
direction, the future railway Gibraltar Strait tunnel and TBM tunnels for HSRL with Maglev
trains, as well as other technological advances that may contribute to improving the construction of TBM traffic tunnels in the future, namely those based on the above mention
concepts.

10.1 MAIN CONCLUSIONS

The history of tunnelling can be said is confused with the history of mankind and its
progress and development over the centuries. Tunnels were first built as human shelter and
later for utilitarian purposes, mainly to transporting water and then integrated into road
networks.

At mid-18th century, with the debut of the (1st) Industrial Revolution, there are a great
increase in the construction of tunnels, following the great development of traffic, with the
emergence of the steam engine and the railway networks. Tunnels were built, essentially on
rocks, using the conventional mining techniques.

The invention of the tunnel shield, patented by Marc Brunel in 1818, allowed the completion,
in 1843, of the first underwater tunnel, the Thames Tunnel, under the Thames River in
London. However, despite this breakthrough and the improvements that were introduced, this
development has not been convincing for a long time, more than a century.

Meanwhile, at the end of the 19th century, other inventions were introduced in the construction of tunnels, such as dynamite and compressed-air drills, as well as the immersed tunnel
technique, which is why different tunnel construction techniques had become available and
have been used, with advantages and disadvantages, depending on the specific conditions of
the tunnels.

In the middle of the 20th century, in 1952, an authentic revolution in the tunnel construction
occurred, when James S. Robbins joined the rotating cutting wheel to the shield, giving rise
to the modern TBMs ("Tunnel Boring Machines"), mechanizing practically all operations,
being today the most common tunneling technique.

Further developments gave rise to different types of TBMs, suitable for the specific conditions of the ground to be bored: Gripper (Open) for hard rock; Single Shield and Double Shield
for soft or fractured rock; EPB ("Earth Pressure Balance") and Slurry Shield (Mixshield)

TBMs for soft soil, and, "the cherry on top of the cake", the multi-mode TBMs that can excavate different types of ground along the same alignment.

Shielded TBMs are authentic mobile factories, which, in addition to excavating the ground, install the precast lining segments that form the tunnel wall, providing frontal and lateral support, which allows excavation under high pressures, in particular from the surrounding water.

Impressive examples of TBM railway and road tunnels have been built around the world in recent times. However, despite their advantages and the breakthroughs achieved, the construction of TBM traffic tunnels still faces significant challenges, namely, appropriate measures to ensure the reliability of the structural safety of tunnels built on soft ground in seismic areas, the most suitable tunnel concept (number of tubes, cross-sectional area of the traffic galleries, etc.), the most appropriate technical systems (ventilation, fire fighting, etc.) and the safety in operation concept, at the lowest cost.

To overcome these challenges, three innovative concepts were recently developed by the author: the TISB concept for TBM tunnels in soft soil in seismic areas, and the TMG and TMF concepts, for TBM railway and road tunnels, respectively.

The TISB concept allows TBM tunnels built on soft ground in seismic areas to be provided with reliable structural safety, with a lower cost than with conventional solutions. The tunnel is formed by two concentric tubes, an outer tube, which is a conventional TBM tunnel, and an inner tube cast in-situ, reinforced with dully confined longitudinal and transverse reinforcements, so as to give the tunnel adequate strength and ductility.

As shown, the application of the TISB concept to the strengthening of a damaged TBM metro tunnel and the construction of a TBM road tunnel allows obtaining tunnels with adequate structural safety with global costs lower than those of conventional solutions.

The TMG and TMF concepts, for railway and road tunnels respectively, allow, with an appropriate lay-out of the tunnel cross-section and the adoption of suitable additional measures, that mono-tube TBM tunnels have the capacities normally offered by two or three parallel tubes connected by cross-passages, thus reducing construction costs, in addition to improving safety in operation, providing the tunnels with innovative and very reliable safety concepts, more reliable than any of the existing ones.

In both cases, three galleries are created inside the tunnel: two traffic galleries (next to each other in the case of railway tunnels or superimposed in the case of road tunnels) and a service (emergency) gallery at the bottom, the three connected by vertical access galleries.

The service gallery serves as a "safe place" for users in incident situations and to provide access to emergency services and the evacuation of users to out of the tunnel, via EMEV ("Emergency Monorail Electric Vehicle") "trains", remotely controlled emergency vehicles, which park at the tunnel portals and circulate suspended on the lower slab. In incident situations, emergency services and users will never make use of the traffic galleries.

As also shown, the application of the TMG and TMF concepts allow for obtaining very cost-effective rail (in particular for HSRL) and road tunnels, with cost reductions of around 20% in rail tunnels and 25% ($1.0 - 0.42/0.56 = 0.25$) in road tunnels, compared to conventional TBM tunnel solutions. This cost reduction, also means less spend of materials, therefore, less "foot print" and impact on nature.

The TISB, TMG and TMF concepts are, therefore, innovations that represent a step forward for the realization of tunnels when using the TBM technique.

10.2 FUTURE PERSPECTIVES

10.2.1 Challenging applications

In addition to the applications presented above, the TISB, TMG and TMF concepts can serve as a basis for carrying out challenging projects in the future, with low cost and high safety in operation.

As paradigmatic examples, the following are the cases of bidirectional TBM mono-tube road tunnels with three lanes in each direction, an alternative solution for the Gibraltar Strait Tunnel, as well as tunnels for High-Speed Railway Lines (HSRL) with Maglev trains.

Road tunnels with 3 lanes in each direction
So far, TBM mono-tube road tunnels for bidirectional traffic with three lanes in each direction have not yet been built.

As mentioned in Chapter 4, ten years ago studies were carried out for a project of this kind, with three lanes of 4.50m clear height, the Orlovski Tunnel, to be held in Saint Petersburg, Russia. Its lay-out required a huge TBM, with 19.25m excavation diameter, a great challenge for TBM industry. However, as mentioned, due to some constraints, the project was cancelled.

However, the adoption of the TMF concept allows the construction of mono-tube TBM road tunnels with three lanes in each direction, with much lower cross-section, even lower than that used in tunnels with two lanes in each direction, as the SR99 Tunnel in Seattle (USA), with 17.50m excavated diameter.

For the inner diameter of the tunnels the author proposes 14.60m, so that, adopting precast segments of 0.60m thick (t ~ Di/25) and a gap of 0.15m to be injected, the excavated diameter of the tunnels will be 16.10m, as shown in figure 10.1. The excavated cross-sectional area of the tunnels is therefore 203 m^2, which represents 70% of the excavated cross-sectional area of the Orlovski Tunnel, with the same capacities (see, Chapter 4).

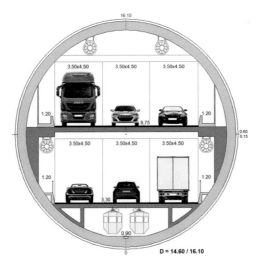

Figure 10.1.   Proposed lay-out for TBM mono-tube road tunnels with 3 lanes in each direction

Inside the tunnels two slabs are built, creating two superimposed road galleries and a service (emergency) gallery at the bottom. Each road gallery houses three lanes with 3.50m width each, making a total road width of 10.50m, and emergency walkways on both sides 1.20m wide. The clear height of the road galleries is 4.50m, so the tunnels present the impressive efficiency index of 0.56.

Above the road lanes there is still 0.50m free space for the signalling installation.

Being based on the TMF concept, the road galleries will be connected to the service gallery below through vertical access galleries, built in one of the sides of the tunnel. Inside the service gallery, dedicated EMEVs will provide access for emergency personnel and the evacuation of users to out of the tunnel in the event of an accident or fire inside the tunnel.

Gibraltar Strait Tunnel

A fixed crossing in the Gibraltar Strait, between Spain and Morroco, is a large project, which has been under study since the 1980's.

After several impressive proposals were made for crossing over a bridge, in 2003 the governments of Spain and Morocco agreed to make the crossing through a railway tunnel.

However, the tunnel has to face great challenges, namely it has to go very deep, more than four hundred meters below sea level, much deeper than any other tunnel existing, and it has to cross ground with bad characteristics.

According to the studies carried out so far under this agreement, the tunnel will link Cape Malabata, near Tanger, North of Marocco, with Punta Paloma, at West of Tarifa, South of Spain. It will be about 39 km long, of which 28 km under the sea, and goes until elevation -412 (figure 10.2) (Pliego, 2005). The maximum speed of the trains will be 120 km/h.

The longitudinal profile of the tunnel foresees two long sections with gradients of 2.5% on the Spanish side and of 2.4% on the Moroccan side, connected by a relatively short almost flat section (figure 10.2).

The geological environment of the site consist mainly of formations called flysches, consisting of clayey materials and sandstone, which are overlaid by post-flysches formations, consisting of shallow granular deposits.

On the marine investigations it was discovered the existence of two very deep parallel paleo-channels (deeper than 600m), in the middle of the strait, about 1 km wide each, mostly composed of deposits of clayey breccia and bioclastic sand, very weak materials. The area is very seismic-prone, and severe earthquakes have been known to occur.

The basic lay-out of the tunnel cross-section was shown on figure 5.1 (see, Chapter 5), which was based on the three-tube tunnel concept, with two unidireccional traffic tubes 7.5m inner diameter and a service tube 4.8m inner diameter, placed in the middle, similar to that of the Channel Tunnel.

Based on TISB and TMG concepts, an alternative solution for the tunnel cross-section was in the meantime developed by the author (Pompeu-Santos, 2008, 2014a, 2014b, 2016) (figure 10.3).

In the alternative solution, the tunnel will be a mono-tube TBM tunnel 11.50m inner diameter inside of which an intermediate concrete slab and a central separating wall are built, creating two independent railway galleries, placed side by side, and a service (emergency) gallery at the bottom (figure 10.3a).

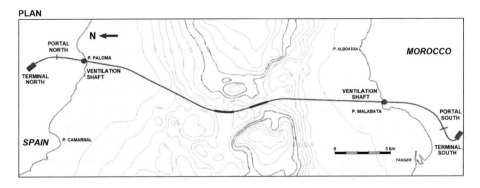

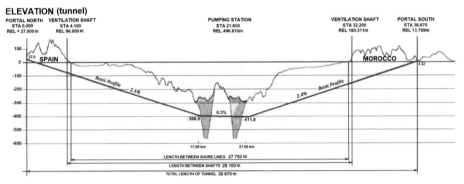

Figure 10.2. Gibraltar Strait Tunnel. Plan and longitudinal section

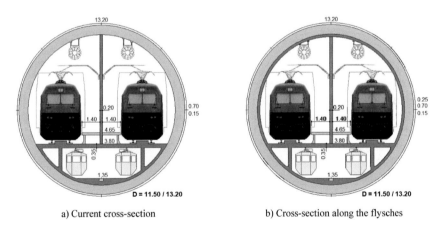

a) Current cross-section   b) Cross-section along the flysches

Figure 10.3. Gibraltar Strait Tunnel. Alternative tunnel cross-section

The cross-sectional area of each railway gallery is about 36 m^2, similar to that of the railway tubes of the basic layout. However, besides the excavated volumes of the two solutions were broadly similar, the alternative solution is much more cost-effective, since, among other advantages, there is no need to build cross-passages.

In the sections of the tunnel where it crosses the paleo-channels it will be strengthened with an interior tube in reinforced concrete 0.25m wall thickness (figure 10.3b), provided with longitudinal and transverse reinforcements dully confined, in order to provide those sections with adequate strength and ductility under earthquakes. The insitu concrete elements that are part of the current cross-section will be part of the insitu tube. The slight reduction in the tunnel cross section is not relevant.

Due to the very high water pressure (42 bar) and the seismic actions, precast segments with 0.70m thickness will be adopted, therefore, considering 0.15m clearance to be injected, the excavated diameter of the tunnel will be 13.20m, however a common value today.

Variant B of the TMG concept will be adopted, so vertical access galleries provide with stairs are arranged in the middle of the tunnel, with escape doors at both sides, spaced about 400m. The galleries next to the service gallery can still be used to install energy lines.

The service gallery will be equipped with emergency vehicles of the EMEV type (see, Chapter 5), to provide local access to the emergency personnel and the evacuation of users to outside in the event of accident or fire inside the tunnel (figure 10.3), as in the other proposed railway tunnels mentioned above.

Given the great length of the tunnel (39 km), the EMEV "trains" will be parked at both tunnel portals. Three cross-overs will also be provided inside the tunnel, thus, large openings about 40m in length, spaced about 10 km apart, and protected with fireproof sliding doors, will be arranged in the separating wall, to allow trains to change the tracks, if necessary.

To be noted that, with this alternative, it would still be possible for the tunnel to be less deep about 80 meters, which would reduce the maximum water pressure to about 3.4 bar and the gradients on both sides of the tunnel to about 2.0%.

Tunnels for HSRL with Maglev trains
As mentioned in Chapter 7, due to the instability of the infrastructure, namely ballast and catenary, conventional High-Speed Railway Lines (HSRL) have their comercial speeds limited to less than 400 km/h.

To go further, other technologies are being tested and some already in use. The currently most promising is the Maglev technology. The "Hyperloop" concept, from which several prototypes have been developed, is still not convincing.

Maglev (derived from "magnetic levitation") is a special system of trains that, instead of using conventional systems of tracks and catenary, uses two sets of magnets: one set to repel and push the train up off the track, and another set to move the elevated train forward, taking advantage of the lack of friction

With Maglev technology, there is just one moving part: the train itself. The train travels along a guideway of magnets which control the train's stability and speed. Propulsion and levitation require no moving parts. This in stark contrast to electric multiple units that may have several dozen parts per bogie. Maglev trains are therefore quieter and smoother than conventional trains and have the potential for much higher speeds.

Although Maglev technology is known from more than one century, it had faced several set-backs, which had impeded a more broad use. Maglev trains are now operational in China, South Korea and Japan. The Shanghai Maglev train also known as the Shanghai Transrapid, has a comercial speed of 430 km/h, the fastest operational high-speed Maglev train, connecting the Shanghai Pudong International Airport and the outskirts of central Pudong, in Shanghai, covering a distance of 30.5 km in 7–8 minutes.

Maglev trains also have the world speed record (604 km/h) and can accelerate and decelerate much faster than conventional high-speed trains.

The power required for levitation does not normally represent a large percentage of the overall energy consumption of a high-speed Maglev system. Overcoming drag, which makes all land transport more energy intensive at high speeds, is what takes the most energy.

Maglev infrastructure is more expensive than conventional train systems, although the simpler construction of Maglev vehicles makes them cheaper to manufacture and maintain.

The main handicap that has been appointed to Maglev trains is that they are not compatible with the conventional infrastructure of HSRL. However, as mentioned, there are already Maglev trains operating in isolated lines and several being tested for longer lines. It is believed that along "medium-long" range routes (500 to 1500 km), Maglev trains will compete favourably with conventional high-speed trains and airplanes

The main Maglev technologies currently in use are the EDS (Electrodynamic suspension) and the EMS (Electromagnetic suspension). However, a Maglev technology called HTS (High-temperature superconducting), recently unveiled (Edwards, 2021), promises even higher speeds than those achieved by EDS and EMS technologies.

The **Electrodynamic suspension (EDS)** uses superconducting electromagnets or strong permanent magnets that create a magnetic field, which induces currents in nearby metallic conductors when there is relative movement, which pushes and pulls the train towards the designed levitation position on the guideway. Both the guideway and the train exert a magnetic field, and the train is levitated by the repulsive and attractive force between these magnetic fields. However, at slow speeds, the resultant magnetic flux is not large enough to levitate the train. Thus, the train must have wheels or some other form of landing gear to support the train until it reaches take-off speed.

This technology is used namely in the experimental SCMaglev train in Japan (the current world speed record holder with 604 km/h) and in the L0 Maglev trains to operate in the future Chuo Shinkansen line (between Tokyo and Nagoya), scheduled for 2027 (figure 10.4a) (JRP, 2019).

a) Electrodynamic suspension (EDS) trains   b) Electromagnetic suspension (EMS) trains

Figure 10.4. Maglev trains

The **Electromagnetic suspension (EMS)** uses electronically controlled electromagnets in the train that attract it to a magnetically conductive (usually steel) track. The train levitates above steel tracks while electromagnets, attached to the train, are oriented toward the tracks below. The system is typically arranged in a series of C-shaped arms, with the upper part of the arms attached to the vehicle and the lower inside edge containing the magnets. The tracks are located inside the Cs, between the upper and lower edges.

This technology is used namely in the Shanghai Transrapid and in the CRRC prototype train that is being developed in China for medium-long distances and speeds of 600 km/h, which is expected to be in operation very soon (figure 10.4b) (Makishuk, 2020).

Tunnels for HSRL with Maglev trains represent a major challenge, however, the TMG concept allows to carry out safe and very cost-benefit solutions. Figure 10.5 presents the tunnel lay-outs for Maglev trains using EDS and EMS technologies proposed by the author, which are based on Variant B of the TMG concept.

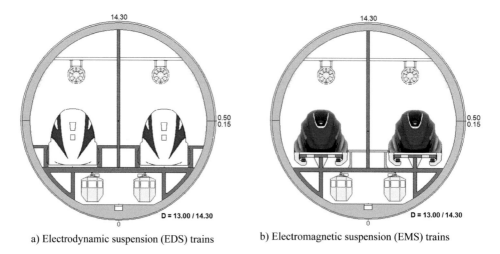

a) Electrodynamic suspension (EDS) trains

b) Electromagnetic suspension (EMS) trains

Figure 10.5.   Proposed lay-out of tunnels for HSRL with Maglev trains

In order to fulfill the safety requirements for so high speeds of the trains, the inner diameter of the tunnels is assumed to be 13.00m, which will provide railway galleries with cross-sectional areas of 56 m² The separating wall will be provided with openings at the ends that act as pressure relieves to provide appropriate comfort conditions inside the trains.

Adopting precast segments of 0.50m thickness (about 1/25 of the inner diameter of the tunnel) and 0.15m clearance to be injected, the excavated diameter of the tunnels will be 14.30m, identical to that of the railway Groeneheart Tunnel, in The Netherlands, presented in Chapter 4.

10.2.2   New technological advances

Another front for the progress in the realization of TBM traffic tunnels and, consequently, those based on the TISB, TMG and TMF concepts are the new technological advances that are being introduced in this activity, namely the robotization of risky operations, the TBM docking and the application of BIM tools in the design and construction of tunnels.

Robotization of risky operations

In TBM tunnelling, some operations present high risks, the most risky being checking and replacing excavation tools (disc cutters, etc.) under pressurized conditions.

Thus, during excavation of the Tuen Mun-Chek Lap Tunnel, innovative remotely operated robotic technology was used, a robotic device called Snake (figure 10.6a). This device incorporates a television camera, high pressure water jet and laser profiling device to measure disc cutter wear. As the name suggests, this device is capable of controlled movement in three dimensions allowing it to access all parts of the cutter head.

The Snake device is eventually used jointly with a disc cutter changing system called Telemach, also remotely controlled, that comprises a disc cutter locking system called Muquet, which was also used in the above mentioned project.

While a typical disc cutter installation comprises eleven parts, Muquet comprises a single piece, which allows the removal and replacement of a single disc cutter to be carried out in less than half an hour, without human intervention at the site.

Another trend is tunnelling at ever greater depths, which brings more and more dangers. In fact, although the technologies outlined above have been successfully proven, the need for entry into a hyperbaric atmosphere remains even if only to recover and repair these tools in the event of a mechanical breakdown or damage.

The construction of the Eurasia Tunnel, for example, was subjected to almost 12 bars of water pressure. Well, a compressed air intervention could not be carried out at these pressure levels without exposing personnel to unacceptable safety hazards.

For these situations, the saturation diving technique is available. This technique comprises a team of divers living in a hyperbaric habitat and being transported to the TBM in a pressurized shuttle (figure 10.6b) that docks with an airlock on the pressure bulkhead of the machine.

a) Snake device   b) Hyperbaric shuttle

Figure 10.6. Robotization of risky operations

The divers breathe mixed gases throughout their stay in the habit on occasion working for months at a time before decompression. Decompression can take a number of days as the pressure is reduced in a series of calibrated steps until atmospheric pressure is achieved.

TBM docking

More and more very long tunnels are being built, particularly railway tunnels, several over 50 km long, and even longer tunnels are being planned, some of them underwater tunnels with more than 100 km in length.

Since excavating long tunnels would take many years, even with modern TBMs, the strategy that has being adopted is the use of several machines, starting from different points along the tunnel.

As mentioned above, these were the cases of the Channel tunnel, 50.5 km long and consisting of three tubes, which was divided in four sections (in which 11 TBMs were used), or the Gothard-Base Tunnel, 57 km long and formed of two tubes, which was divided in five sections (in which 4 TBMs were used).

Therefore, horizontal access galleries were built along the tunnels, from where the TBMs started to excavate until to achieve the next access point, where they were disassembled and put out of the tunnel or moved to another access point to start excavating another section.

However, in some situations, namely in underwater tunnels, this strategy can no longer be adopted as no access shafts or galleries are possible to be built under the water. To overcome this situation, two boring machines are usually used, starting from the tunnel portals or access shafts in-shore, which will meet halfway, accomplishing their mission.

The recovery of the rear backups of the machines usually does not put particular difficulties, as they are made of pieces not very heavy and mostly connected by bolts. Regarding the shields, which enclose the most sophisticated components of the machine, things are more complicated. In this situation different strategies may be adopted.

One solution is to bury the shield of one of the TBMs inside the ground, with the loss of the equipment. This was the solution adopted in the Channel Tunnel in the English side, in which, as mentioned, the shields of the three TBMs that started from the access shafts on the English margin were buried in the middle of the Channel.

Another solution is to create large caverns in the locations where the TBMs meet, inside which they are disassembled in pieces of adequate weight and size, from where they are transported to the tunnel portals, as happened with the TBMs on the French side of the referred tunnel, or the Gothard-Base Tunnel, as also mentioned. However, caverns are expensive to build and require ground with appropriate waterproof characteristics, which sometimes don't exist.

An alternative promising solution is the adoption of the so called TBM docking. In the TBM docking, the machines advance against each other and when the first reaches the meeting point it stops, while the other continues until they are face to face, in order to dock. However, few underground TBM docking methods have been tried, so far.

In the conceptual docking method, the shield of the penetrating (arriving) TBM is provided with a steel ring that penetrates into a pressure-receiving rubber ring in the accepting (waiting) TBM, so that the two machines dock to each other mechanically, like satellites and space shuttles (figure 10.7a).

This operation, however, presents some difficulties, namely regarding the alignment of the shields and the guarantee of the tightness of the connection.

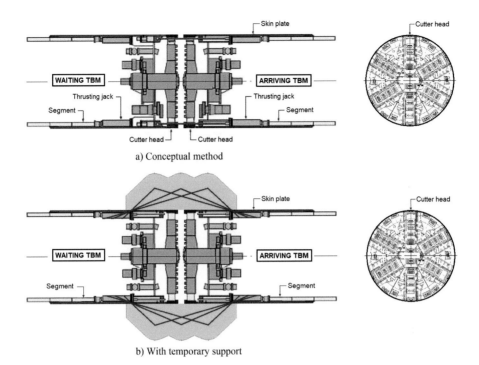

a) Conceptual method

b) With temporary support

Figure 10.7. TBM docking

A simplified solution can however be used, in which the cutter heads of the machines do not touch each other, the ground between them being excavated after the execution of a temporary support made of injections of jet grouting or resin grout, or by the local freezing of the ground (figure 10.7b). The skin plates of the shields of the TBMs will also be part of the temporary support system.

The machine components inside the shields are then disassembled and transported out of the tunnel. The excavation is carried out afterwards and the internal lining is executed inside the shields, filling the space between the precast segments placed by the two TBMs.

For the feasibility of the docking method easy disassembling TBMs are essential. Therefore, further progress is needed, in order these operations become trivial.

Considering the excavation rates offered by the existing TBMs (more than 200m per week in rock), the adoption of the docking method will make feasible to bore underwater tunnels over 100 km long without using intermediate shafts within a reasonable time, 5–6 years.

This may be the case of the FinEst Tunnel, in the Gulf of Finland, between Helsinki (Finland) and Tallinn (Estonia), 102 km long, or the Bohai Strait Tunnel, in northern China, 106 km long, which they are now in the planning stage, where also the application of the TMG concept can be of great benefit, both in terms of safety and cost.

Application of BIM tools

BIM (Building Information Modelling) provides a digital representation of physical and functional characteristics of a facility, which can easily be communicated with technical and non-technical stakeholders, so that appropriate decisions on its life-cycle can be made.

BIM is a relatively new paradigm in the construction industry envisaging to integrate three pillars: people, process and technology to deliver assets that meet client's requirements (Khontothanasis P et al. (2020). The key word is interoperability.

Despite its conservative character, the construction industry has been able to meet a lot of challenges thanks to digital innovations and enhanced interoperability. With the advent of Industry 4.0 (4th Industrial Revolution), there is an increased focus on delivering smarter construction and efficient operations and maintenance. This can include better use of networks such as 5G, smart buildings, IoT (Internet of the Things) and sensor data.

The application of BIM tools in tunnelling engineering, in particular TBM tunnels, will ensure consistency among stakeholders in different maturity level during the project management life-cycle, contributing to improved solutions, with benefits for all.

To start a project, a BIM management team should be set up, however, cooperation between the teams involved is instrumental to the success of the project. A common data environment for graphical and non-graphical data and documents has to be established.

A 3D model of the structure is firstly created, usually from the geo-referenced tunnel alignment, which becomes the information base for the definition of the construction elements, namely the 2D cross-sections, or the lay-out of pipes and cables. BIM assures a complete consistency of the three-dimensional representation of all components. Depending on the purpose, information with different LoDs (Levels of Detail) can be used (figure 10.8) (Ninic, 2017).

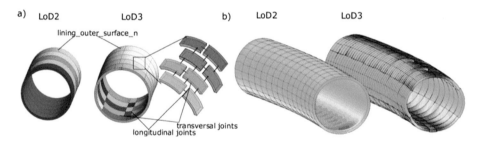

Figure 10.8.   Modelling the behaviour of a TBM tunnel lining with BIM for two LoDs (courtesy of J. Ninic)

Below, other aspects are introduced: 4D (Time), 5D (Costs), 6D (Sustainability) and 7D (Operation), with a view to an effective project management throughout the complete life-cycle of the tunnel.

4D will be very useful for tunnels based on the TISB, TMG and TMF concepts, as it will allow to optimize the sequence of the complementary works to be carried out inside the tunnel (Execution Plan).

Other areas of improvement

In addition to the increasing use of multi-mode TBMs and accessible cutter heads, mentioned above, other technological improvements in TBM tunnelling include:

– **Collecting data**: Modern TBMs can collect and process vast amounts of data in real time, which allows, for example, very close control of ground movements, permitting the construction of tunnels in dense urban environments without damage to buildings or subsurface utilities.

– **Inspection automation**: Deployment of cameras in the excavating chamber, as well as electronic sensors in the cutting tools, will allow repairs or exchanges to be scheduled, leading to better performance of the TBMs.

– **Use of robotics**: The use of robots in risky operations (such as changing cutting tools) or repetitive operations (such as drilling holes) will allow the faster construction of tunnels.

– **Advance geological survey**: Advance surveys, using techniques such as seismic exploration, probe drilling or penetrating radar, allow to identification of the location of boulders or other obstructions in the front, reducing the risk of unplanned TBM stops.

– **Improving additives**: Increased scientific knowledge about additives to be added into excavated soil in the pressurized area at the front of the TBM, allow for providing reliable uniform support to the ground around the tunnel.

– **Improving seals**: Improved seals in bearings and between the TBM shield and the tunnel lining allow tunnels to be built under increasing groundwater pressures.

– **Re-use of TBMs**: Re-employment of machines and components shall be a way to go, as it will provide a substantial reduction in environment impact; will save material and energy and improve the carbon footprint of TBMs, thus contributing to sustainable tunnelling.

Automation of the TBMs

The ultimate goal is the automation of the TBMs (autonomous TBMs), so that they can operate entirely on their own, without the need for human operators at site, which will increase efficiency (and safety) in the construction of the tunnels in the future.

10.3   FINAL REMARKS

Tunnels are playing an increasingly major role in building transport infrastructure worldwide. Titles like: "Underground, the way to the future", "The future goes underground" or "Tunnels for a better life" are not just marketing; they correspond to reality.

Overcoming the recent international shortages (Covid19 pandemic, energy crisis associated with the war in Ukraine, etc.), major investments will be made in infrastructure around the world, especially in rail and road networks. Tunnel construction will play a key role, also with a view to contribute to a greener environment. So, with the 20th century being mainly focused on bridges, it is to be expected that the 21th century will the time of tunnels.

Great progress has been made in the last two decades in the construction of tunnels, in particular with the TBM technique, with the adoption of outstanding innovation, allowing for increasingly economical and safe tunnels. However, further improvements are to be expected in the near future.

As shown, the TISB, TMG and TMF concepts, recently developed by the author, allow for the obtaining TBM traffic tunnels even more economical and with greater safety in operation, being the right approach for the realization of more and more challenging traffic tunnels. On the other hand, advances in TBM technology are in continuous progress, allowing the realization of better tunnels and in safer conditions, and more sustainable.

In summary, it can be concluded that, in building a better world, tunnelling, in particular with the TBM technique, has a bright future ahead.

REFERENCES

Arup (2015), *Cost Benefits of Large Diameter Bored Tunnels*. TunnelTalk, April 2015

Duhme R (2017), *Designing TBMs for Subsea Tunnels*. The 2017 World Congress on Advances in Structural Engineering and Mechanics, Ilsan, Korea, August 2017

Edwards C (2021), *China Unveils 400mph Levitating "Magnet Train" that's Nearly as Fast as Jet Plane*. The Sun, January 14, 2021

Frew B (2020), *Advancements in Large Diameter TBM Tunnelling*. Tunnel Business Magazine, February 21, 2020

JRP (2019), *The Japanese Maglev: World's Fastest Bullet Train*. Japan Rail Press, November 15, 2019

Khontothanasis P et al. (2020), *BIM and Advanced Computer-Based Tools for the Design and Construction of Underground Structures and Tunnels*. Tunnel Engineering – Selected Topics (Chapter 10), IntechOpen, London, UK, April 2020

Makishuk D (2020), *China "Floating" Maglev Train in Testing Stage*. Asia Times, June 23, 2020

Ninic J (2017), *Parametric Multi-Level Tunnel Modelling for Design Support and Analysis*. IV International Conference on Computational Methods in Tunnelling and Subsurface Engineering (EURO: TUN2017), Innsbruck, Austria, April 2017

Pliego J M (2005), *Opening Session – Briefing Document*. UN-ITA Workshop Systematic Ground Probing and Treatment in Mechanised Tunnelling, Madrid, Spain, January 2005.

Pompeu-Santos S (2008), *The TISB and TMG Concepts. Application on a Proposal for the Railway Tunnel through the Gibraltar Strait*. IABSE Congress Creating and Renewing Urban Structures: Tall Buildings, Bridges and Infrastructure, Chicago, USA, September 2008

Pompeu-Santos S (2012) *Tunnels for High-Speed Railway Lines Based on the TMG Concept*. IABSE Congress Innovative Infrastructures – Toward Human Urbanism, Seoul, Korea, September 2012

Pompeu-Santos S (2014a), *High-Speed Railway Tunnels Based on the TMG Concept*. World Tunnel Congress (WTC2014) Tunnels for a Better Life, Foz do Iguassu, Brazil, May 2014

Pompeu-Santos S (2014b), *New Trends in the Tunnels of the Future*. IABSE Symposium Engineering for Progress, Nature and People, Madrid, Spain, September 2014

Pompeu-Santos S (2016), *Sustainable TBM Tunnels for Tomorrow*. Second International Conference on Concrete Sustainability (ICCS16), Madrid, Spain, June 2016

Senthilnath G (2016), *Mid-Tunnel Underground Docking of Tunnel Boring Machines in Singapore*. www.researchgate.net, January 2016

TM-CLKL (2021), *Tuen Mun-Check Lap Kok Link. Engineering Features*. https://tm-clkl.hk/eng/achievements_tmclkl.html, 2021

Willis D et al (2020), *Sustainable TBM Design for Long-Term Usage*. Tunnel, N° 2, 2020

Wikipedia (2021), *Maglev*. Wikipedia, 2021

Index